GCSE
Mathematics

At CGP, we're a bit obsessed with GCSE Maths... so we had a great time packing this book with everything you need for the Foundation Level exams. It's a Maths extravaganza.

As you'd expect, we've included unbeatably clear notes and examples, plus brilliant practice questions and realistic exam papers. And don't forget the amazing digital extras — there's a free Online Edition, interactive quizzes and step-by-step video solutions.

Once you've got through all that, you'll love GCSE Maths almost as much as we do.

Unlock your free online extras!

Just go to **cgpbooks.co.uk/extras** and enter this code or scan the QR codes in the book.

2899 6026 9625 7558

By the way, this code only works for one person. If somebody else has used this book before you, they might have already claimed the Online Edition.

Complete
Revision & Practice

Everything you need to pass the exams!

Contents

Throughout this book you'll see **grade stamps** like these:

You can use these to focus your revision on easier or harder work.

But remember — to get a top grade you have to know **everything**, not just the hardest topics.

You'll see **QR codes** throughout the book that you can scan with your smartphone.

A QR code next to an exam-style question takes you to a **video** that talks you through solving the question. You can access **all** the videos by scanning this code here.

A QR code on a 'Revision Questions' page takes you to a **Sudden Fail Quiz** for that topic. Fancy trying a quiz covering **every topic** in the book? Then scan this code right here!

You can also find the **full set of videos** at cgpbooks.co.uk/GCSEMathsVideos-Foundation and the **full set of quizzes** at cgpbooks.co.uk/GCSEMathsQuiz-Foundation

Published by CGP

From original material by Richard Parsons

Updated by: Martha Bozic, Eleanor Crabtree, Sammy El-Bahrawy, Sarah George,
    Ruth Greenhalgh, David Ryan, Caley Simpson

Contributor: Alastair Duncombe

With thanks to Simon Little and Michael Weynberg for the proofreading

With thanks to Jan Greenway for the copyright research

Printed by Elanders Ltd, Newcastle upon Tyne.
Clipart from Corel®

Types of Number and BODMAS

Ah, the glorious world of GCSE Maths. Get stuck into this first section and you'll be an expert in no time. Here are some handy definitions of different types of number, and a bit about what order to do things in.

Integers

An <u>integer</u> is another name for a <u>whole number</u> — either a positive or negative number, or zero.

<u>Examples</u>

Integers:	-365, 0, 1, 17, 989, $1\,234\,567\,890$
Not integers:	0.5, $\frac{2}{3}$, $\sqrt{7}$, $13\frac{3}{4}$, -1000.1, 66.66, π

Square and Cube Numbers

1) When you <u>multiply</u> a whole number by <u>itself</u>, you get a <u>square number</u>:

1^2	2^2	3^2	4^2	5^2	6^2	7^2	8^2	9^2	10^2	11^2	12^2	13^2	14^2	15^2
1	4	9	16	25	36	49	64	81	100	121	144	169	196	225
(1×1)	(2×2)	(3×3)	(4×4)	(5×5)	(6×6)	(7×7)	(8×8)	(9×9)	(10×10)	(11×11)	(12×12)	(13×13)	(14×14)	(15×15)

2) When you <u>multiply</u> a whole number by <u>itself</u>, then by itself <u>again</u>, you get a <u>cube number</u>:

1^3	2^3	3^3	4^3	5^3	10^3
1	8	27	64	125	1000
(1×1×1)	(2×2×2)	(3×3×3)	(4×4×4)	(5×5×5)	(10×10×10)

3) You should know these basic squares and cubes <u>by heart</u> — they could come up on a non-calculator paper, so it'll save you time if you already know what they are.

BODMAS <u>B</u>rackets, <u>O</u>ther, <u>D</u>ivision, <u>M</u>ultiplication, <u>A</u>ddition, <u>S</u>ubtraction

<u>BODMAS</u> tells you the <u>ORDER</u> in which these operations should be done:
Work out <u>Brackets</u> first, then <u>Other</u> things like squaring, then <u>Divide</u> / <u>Multiply</u> groups of numbers before <u>Adding</u> or <u>Subtracting</u> them.

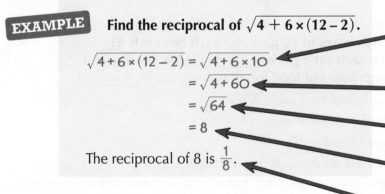

EXAMPLE **Find the reciprocal of $\sqrt{4 + 6 \times (12 - 2)}$.**

$$\sqrt{4 + 6 \times (12 - 2)} = \sqrt{4 + 6 \times 10}$$
$$= \sqrt{4 + 60}$$
$$= \sqrt{64}$$
$$= 8$$

The reciprocal of 8 is $\frac{1}{8}$.

It's not obvious what to do inside the square root — so use BODMAS. <u>Brackets</u> first...

... then <u>multiply</u>...

... then <u>add</u>.

<u>Take the square root</u>

Finally, take the <u>reciprocal</u> (the reciprocal of a number is just $1 \div$ the number).

Brackets, Other, Division, Multiplication, Addition, Subtraction

It's really important to check your working on BODMAS questions. You might be certain you did it right, but it's surprisingly easy to make a slip. Try this Exam Practice Question and see how you do:

Q1   Without using a calculator, find the value of $3 + 22 \times 3 - 14$.   [2 marks]

Wordy Real-Life Problems

*Wordy questions can be a bit off-putting — for these you don't just have to do **the maths**, you've got to work out what the question's **asking you to do**. It's **really** important that you begin by reading through the question **really** carefully...*

Don't Be Scared of **Wordy Questions**

<u>Relax</u> and work through them <u>step by step</u>:

> 1) <u>READ</u> the question <u>carefully</u>. Work out <u>what bit of maths</u> you need to answer it.
> 2) <u>Underline</u> the <u>INFORMATION YOU NEED</u> to answer the question — you might not have to use <u>all</u> the numbers they give you.
> 3) Write out the question <u>IN MATHS</u> and answer it, showing all your <u>working</u> clearly.

 EXAMPLES

1.   **A return car journey from Carlisle to Manchester uses $\frac{4}{7}$ of a tank of petrol. It costs <u>£56</u> for a <u>full tank</u> of petrol. How much does the journey cost?**

You want to know $\frac{4}{7}$ of £56, so in maths: $\frac{4}{7}$ × £56 = £32

See p.15-17 for more on using fractions.

Don't forget the units in your final answer — this is a question about <u>cost in pounds</u>, so the units will be <u>£</u>.

2.   **A water company has two different price plans for water usage:**

PLAN A	PLAN B
No fixed cost	**Fixed cost of £12**
23p per unit	**Plus 11p per unit**

A household uses <u>102 units</u> of water in one month. Which price plan would be cheaper for this household in that particular month?

Cost with plan A: 102 × £0.23 = £23.46

Cost with plan B: £12 + (102 × £0.11) = £23.22

So for this household in that particular month, plan B would be cheaper.

3.   **Vashti buys dog food in boxes of <u>20 packets</u>. Each box costs <u>£12.50</u>. She has <u>3 dogs</u> which each eat <u>2 packets per day</u>. How much will it cost her to buy enough boxes of food for all of her dogs for <u>4 weeks</u>?**

Number of packets for 3 dogs for 1 day = 3 × 2 = 6

Number of packets for 3 dogs for 4 weeks = 6 × 7 × 4 = 168

Number of boxes needed = 168 ÷ 20 = 8.4

Vashti can't buy part of a box, so she needs to buy 9 boxes.

9 × £12.50 = £112.50

Just because it's wordy doesn't mean it's difficult...

A lot of these wordy questions have a few stages to the working — if you take them one step at a time, they're really not bad at all. Have a go at breaking down this Exam Practice Question.

Q1   Hank owns a fruit stall. He sells 11 apples and 5 oranges for £2.71. One apple costs 16p. How much does one orange cost?   [4 marks]   (3)GRADE

Q1 Video Solution

Multiplying and Dividing by 10, 100, etc.

*You need to know the stuff on this page — it's quite simple and they're likely to **test you on it** in the exam.*

1) **Multiplying** and **Dividing** Any Number by **10**

Multiply

Move the decimal point <u>ONE</u> place <u>BIGGER</u> ( ) and if it's needed, <u>ADD A ZERO</u> on the end.

E.g. 23.6 × 10 = 2 3 6

Divide

Move the decimal point <u>ONE</u> place <u>SMALLER</u> ( ) and if it's needed, <u>REMOVE ZEROS</u> after the decimal point.

E.g. 23.6 ÷ 10 = 2 . 3 6

2) **Multiplying** and **Dividing** Any Number by **100**

Multiply

Move the decimal point <u>TWO</u> places <u>BIGGER</u> and <u>ADD ZEROS</u> if necessary.

E.g. 34 × 100 = 3 4 0 0

Divide

Move the decimal point <u>TWO</u> places <u>SMALLER</u> and <u>REMOVE ZEROS</u> after the decimal point.

E.g. 340 ÷ 100 = 3 . 4

3) **Multiplying** and **Dividing** by **1000** or **10 000**

Multiply

Move the decimal point so many places <u>BIGGER</u> and <u>ADD ZEROS</u> if necessary.

Divide

Move the decimal point so many places <u>SMALLER</u> and <u>REMOVE ZEROS</u> after the decimal point.

You always <u>move</u> the <u>decimal point</u> this much:
<u>1 place for 10,</u>   <u>2 places for 100,</u>
<u>3 places for 1000,</u>   <u>4 places for 10 000,</u>   etc.

4) **Multiply** and **Divide** by Numbers like **20, 300**, etc.

<u>MULTIPLY</u> by <u>2</u> or <u>3</u> or <u>8</u> etc. <u>FIRST</u>, then move the decimal point so many places <u>BIGGER</u>, according to how many zeros there are.

<u>DIVIDE</u> by <u>4</u> or <u>3</u> or <u>7</u> etc. <u>FIRST</u>, then move the decimal point so many places <u>SMALLER</u>, according to how many zeros there are.

EXAMPLES

1. **Calculate 234 × 200.**

234 × 2 = 468
468 × 100 = 46800

2. **Calculate 960 ÷ 300.**

960 ÷ 3 = 320
320 ÷ 100 = 3.2

Multiplying — move the decimal point to the right

When dividing move the decimal point to the left. Seems easy, but still needs practice, so try these:

Q1   Work out   a) 12.3 × 100   b) 3.08 ÷ 1000   c) 360 ÷ 30   [3 marks]

Multiplying and Dividing Whole Numbers

*You need to be really happy doing multiplications and divisions **without** a calculator
— they're likely to come up in your **non-calculator** exam.*

Multiplying **Whole Numbers**

The Traditional Method:

1) Split it into <u>separate multiplications</u>.
2) Add up the results in <u>columns</u> (right to left).

 1. Work out 46 × 27

```
      4 6
    × 2 7
    ─────
    3 2₄2  —— This is 7 × 46
    9,2 0  —— This is 20 × 46
    ─────
    1 2 4 2  —— This is 322 + 920
```

2. **Work out 243 × 18**

```
      2 4 3
    ×   1 8
    ───────
    1 9₃4₂4  —— This is 8 × 243
    2 4 3 0  —— This is 10 × 243
    ───────
    4₁3 7 4  —— This is 1944 + 2430
```

There are lots of other multiplication methods — make sure you're comfortable using whichever method you prefer.

Dividing **Whole Numbers**

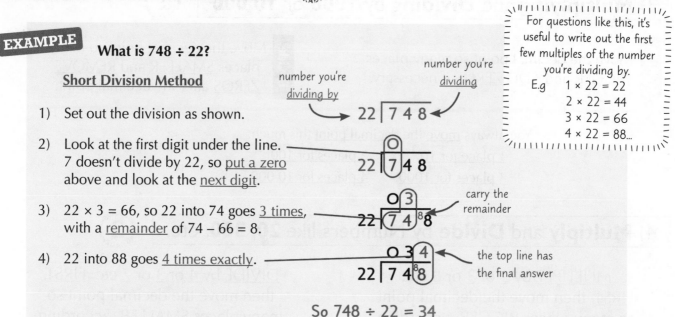

EXAMPLE

What is 748 ÷ 22?

Short Division Method

number you're <u>dividing by</u>

number you're <u>dividing</u>

1) Set out the division as shown. → 22 | 7 4 8 ←

For questions like this, it's useful to write out the first few multiples of the number you're dividing by.
E.g   1 × 22 = 22
       2 × 22 = 44
       3 × 22 = 66
       4 × 22 = 88...

2) Look at the first digit under the line. 7 doesn't divide by 22, so put a zero above and look at the <u>next digit</u>.

    0
22 | 7 4 8

3) 22 × 3 = 66, so 22 into 74 goes <u>3 times</u>, with a <u>remainder</u> of 74 − 66 = 8.

    0 3
22 | 7 4 ⁸8   carry the remainder

4) 22 into 88 goes <u>4 times exactly</u>.

    0 3 4
22 | 7 4 ⁸8   the top line has the final answer

So 748 ÷ 22 = 34

The other common method for dividing is <u>long division</u> — if you prefer this method, make sure you know it <u>really</u> well, so you'll have no problems with any division in your exam.

Make sure you know how to multiply and divide without a calculator

Practice makes perfect when it comes to multiplication and division. Go through the worked examples on this page and make sure you can follow the methods, then try these practice questions... <u>without</u> a calculator.

Q1 Work out a) 28 × 12   b) 56 × 11   c) 104 × 16   [6 marks]

Q2 Work out a) 96 ÷ 8   b) 84 ÷ 7   c) 252 ÷ 12   [4 marks]

Q3 Joey has a plank of wood which is 220 cm long. He cuts it into 14 cm pieces. What length of wood will he have left over?   [2 marks]

Q3 Video Solution

Multiplying and Dividing with Decimals

You might get a scary non-calculator question on multiplying or dividing using decimals. Luckily, these aren't really any harder than the whole-number versions. You just need to know what to do in each case.

Multiplying Decimals

1) Start by <u>ignoring</u> the decimal points. Do the multiplication using <u>whole numbers</u>.
2) Count the <u>total</u> number of digits after the <u>decimal points</u> in the original numbers.
3) Make the answer have the <u>same number</u> of decimal places.

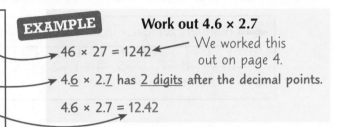

EXAMPLE **Work out 4.6 × 2.7**

46 × 27 = 1242 ← We worked this out on page 4.

4.6 × 2.7 has 2 digits after the decimal points.

4.6 × 2.7 = 12.42

Dividing a **Decimal** by a **Whole Number**

For these, you just set the question out like a whole-number division <u>but</u> put the <u>decimal point</u> in the answer <u>right above</u> the one in the question.

EXAMPLE **What is 52.8 ÷ 3?** Put the decimal point in the answer above the one in the question.

```
   1 .              1 7 .            1 7 . 6
3 | 5 ²2 . 8     3 | 5 ²2 .¹8     3 | 5 ²2 .¹8    So 52.8 ÷ 3 = 17.6
```

3 into 5 goes once, carry the remainder of 2.

3 into 22 goes 7 times, carry the remainder of 1.

3 into 18 goes 6 times exactly.

Dividing a **Number** by a **Decimal**

Two-for-one here — this works if you're dividing a whole number by a decimal, or a decimal by a decimal.

EXAMPLE **What is 36.6 ÷ 0.12?**

1) The trick here is to write it as a fraction:

$$36.6 ÷ 0.12 = \frac{36.6}{0.12}$$

2) Get rid of the decimals by multiplying top and bottom by 100 (see p.3):

$$= \frac{3660}{12}$$

3) It's now a decimal-free division that you know how to solve:

```
        0              0 3             0 3 0            0 3 0 5
12 | 3 ³6 6 0    12 | 3 ³6 6 0    12 | 3 ³6 6⁶0    12 | 3 ³6 6⁶0    So 36.6 ÷ 0.12 = 305
```

12 into 3 won't go so carry the 3

12 into 36 goes 3 times exactly

12 into 6 won't go so carry the 6

12 into 60 goes 5 times exactly

To divide decimals by decimals, first turn them into whole numbers

Just be careful to multiply each decimal by the same amount when you do. Have a go at these questions:

Q1 Work out a) 3.2 × 56 b) 0.6 × 10.2 c) 5.5 × 10.2 [6 marks]

Q2 Emma buys 14 packets of biscuits for a party. Each packet of biscuits costs £1.27. How much does Emma spend on biscuits in total? [2 marks]

Q3 Claire has £22.38. She wants to share it equally between her three nephews. How much does each nephew receive? [2 marks]

Q4 Find: a) 33.6 ÷ 0.6 b) 45 ÷ 1.5 c) 84.6 ÷ 0.12 [6 marks]

Negative Numbers

*Numbers less than zero are **negative**. You should be able to **add**, **subtract**, **multiply** and **divide** with them.*

Adding and Subtracting with Negative Numbers

Use the number line for addition and subtraction involving negative numbers:

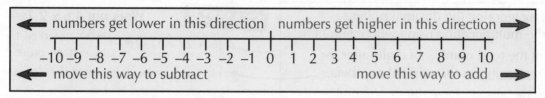

← numbers get lower in this direction   numbers get higher in this direction →

–10 –9 –8 –7 –6 –5 –4 –3 –2 –1 0 1 2 3 4 5 6 7 8 9 10

← move this way to subtract   move this way to add →

EXAMPLES

What is –4 + 7?   Start at –4 and move 7 places in the positive direction:

–5 –4 –3 –2 –1 O 1 2 3 4   So –4 + 7 = 3

Work out 5 – 8   Start at 5 and move 8 places in the negative direction:

–4 –3 –2 –1 O 1 2 3 4 5 6   So 5 – 8 = –3

Find –2 – 4   Start at –2 and move 4 places in the negative direction:

–6 –5 –4 –3 –2 –1 O   So –2 – 4 = –6

Use These Rules for Combining Signs

+	+	makes	+
+	–	makes	–
–	+	makes	–
–	–	makes	+

These rules are ONLY TO BE USED WHEN:

1) Multiplying or dividing

EXAMPLES

Find: a) –2 × 3 ⌒(invisible + sign)   – + makes – so –2 × 3 = –6

b) –8 ÷ –2   – – makes + so –8 ÷ –2 = 4

2) Two signs appear next to each other

EXAMPLES

Work out: a) 5 – –4   – – makes + so 5 – –4 = 5 + 4 = 9

b) 4 + –6 – –7   + – makes – and – – makes +

so 4 + –6 – –7 = 4 – 6 + 7 = 5

> Be careful when squaring or cubing. Squaring a negative number gives a positive number (e.g. –2 × –2 = 4) but cubing a negative number gives a negative number (e.g. –3 × –3 × –3 = –27).

Number lines are handy for adding or subtracting negative numbers

Don't just learn the rules in that burgundy box — make sure you know when you can use them too.

Q1   The temperature in Mathchester at 9 am on Monday was 4 °C.
At 9 am on Tuesday the temperature was –2 °C.
a) What was the change in temperature from Monday to Tuesday?   [1 mark]
b) The temperature at 9 am on Wednesday was 3 °C lower than on Tuesday.
What was the temperature on Wednesday?   [1 mark]

Warm-up and Worked Exam Questions

Doing maths without a calculator becomes easier the more you practise. These warm-up questions will help to get your brain in gear. Work through them without using your calculator.

Warm-up Questions

1) Without using a calculator, work out $7 + 15 - 6 \div 3$.

2) A ticket for entry to a funfair plus 3 rides costs £2.60. After that, each extra ride costs 95p. Roshani and Eve go on 8 rides each, how much do they pay altogether?

3) Work out: a) 0.957×1000 b) 2.4×20

4) Work out: a) $2.45 \div 10$ b) $4000 \div 800$

5) Work out the following:
   a) 43×9 b) 32×17 c) 5.8×21 d) 1.3×6.5

6) Work out the following:
   a) $91 \div 7$ b) $242 \div 2$ c) $24.8 \div 0.4$ d) $36 \div 1.2$

7) Work out: a) -4×-3 b) $-4 + -5 + 3$ c) $(3 + -2 - 4) \times (2 + -5)$ d) $120 \div -40$

Worked Exam Questions

These questions already have the answers filled in. Have a careful read through the working and handy hints before you have a go at the exam questions on the next page.

1   Yonas has a 500 ml bottle of a fizzy drink. Poppy has 216 ml of the same fizzy drink in a glass. Yonas gives Poppy some of his drink so that they each have the same amount.

How much drink does Yonas give to Poppy?

Find how much fizzy drink there is in total.

Total amount of drink = 500 ml + 216 ml = 716 ml

$$2)\overline{7^11^16} = 358 \text{ ml each}$$
$$358$$

Divide this by 2 to find the amount they should each have.

500 ml − 358 ml = 142 ml,

so Yonas gives 142 ml of his drink to Poppy.

...................142........... ml
[2 marks]

2   Work out how many of each item below were sold if: **3**

a)   Tamera spent £14 on rulers that cost £0.70 each.

$$14 \div 0.7 = \frac{14}{0.7} = \frac{140}{7}$$

$$7)\overline{1^14\,0}$$
$$0\,2\,0$$

Multiply top and bottom by 10, then do the division as normal.

...................20...........
[2 marks]

b)   Michael spent £2.76 on pencils that cost £0.12 each.

$$2.76 \div 0.12 = \frac{2.76}{0.12} = \frac{276}{12}$$

$$12)\overline{2^27^36}$$
$$0\,2\,3$$

Don't worry if you have to keep carrying numbers — just continue like normal.

...................23...........
[2 marks]

Exam Questions

3 Write down the value of each of the following:

    a) 9^2 **(1)**

....
[1 mark]

    b) 4^3 **(2)**

....
[1 mark]

4 Three numbers multiply to give 288. Two of the numbers are –3 and 12. **(2)**

    What is the third number?

....
[3 marks]

5 Alanna buys 15 tickets for a concert for herself and some friends. **(3)**
    Each ticket is the same price. She pays with £200 and gets £5 change.

    How much does each ticket cost?

£ ....
[3 marks]

6 Georgie is a sales representative. She drives to **(4)**
    different companies to sell air conditioning units.

    When she has to travel, her employer pays fuel expenses of 30p per mile.
    She drives to a job in the morning and drives home again later that day.
    She is also given £8 to cover any food expenses for each day that she is not in the office.

    The distances to her jobs for this
    week are shown on the right.

    Find Georgie's total expenses for this week.

Hint: think carefully
about the total
distance travelled.

Monday: Buckshaw, 30 miles
Tuesday: in office
Wednesday: Wortham, 28 miles
Thursday: Harborough, 39 miles
Friday: Scotby, 40 miles

£ ....
[4 marks]

Prime Numbers

There's no getting around it — prime numbers are as important as they sound.
Luckily, they're not too difficult.

PRIME Numbers **Don't Divide** by Anything

Prime numbers are all the numbers that DON'T come up in times tables:

| 2 | 3 | 5 | 7 | 11 | 13 | 17 | 19 | 23 | 29 | 31 | 37 | ... |

The only way to get ANY PRIME NUMBER is: 1 × ITSELF

E.g.   The only numbers that multiply to give 7 are 1 × 7
        The only numbers that multiply to give 31 are 1 × 31

> **EXAMPLE**   **Show that 24 is not a prime number.**
>
> Just find another way to make 24 other than 1 × 24:   2 × 12 = 24
>
> 24 divides by numbers other than 1 and 24, so it isn't a prime number.

Five **Important Facts**

1) 1 is NOT a prime number.

2) 2 is the ONLY even prime number.

3) The first four prime numbers are 2, 3, 5 and 7.

4) Prime numbers end in 1, 3, 7 or 9 (2 and 5 are the only exceptions to this rule).

5) But NOT ALL numbers ending in 1, 3, 7 or 9 are primes, as shown here:
   (Only the circled ones are primes.)

How to **FIND** Prime Numbers — a very simple method

> 1) All primes (above 5) end in 1, 3, 7 or 9. So ignore any numbers that don't end in one of those.
>
> 2) Now, to find which of them ACTUALLY ARE primes you only need to divide each one by 3 and by 7. If it doesn't divide exactly by 3 or by 7 then it's a prime.

This simple rule using just 3 and 7 is true for checking primes up to 120.

> **EXAMPLE**   **Find all the prime numbers in this list: 71, 72, 73, 74, 75, 76, 77, 78**
>
> ① First, get rid of anything that doesn't end in 1, 3, 7 or 9: 71, 7̶2̶, 73, 7̶4̶, 7̶5̶, 7̶6̶, 77, 7̶8̶
>
> ② Now try dividing 71, 73 and 77 by 3 and 7:
>
> $71 \div 3 = 23.667$ $71 \div 7 = 10.143$ so 71 is a prime number
>
> $73 \div 3 = 24.333$ $73 \div 7 = 10.429$ so 73 is a prime number
>
> $77 \div 3 = 25.667$ BUT: $77 \div 7 = 11$ — 11 is a whole number,
> so 77 is NOT a prime, because it divides by 7.
>
> So the prime numbers in the list are 71 and 73.

Remember — prime numbers don't come up in times tables

Learn all three sections above, then cover the page and try this Exam Practice Question without looking:

Q1 Below is a list of numbers. Write down all the prime numbers from the list.
39, 51, 46, 35, 61, 53, 42, 47   [2 marks]

Multiples, Factors and Prime Factors

You might get asked to list multiples or find factors in the exam. This page will teach you how.

Multiples and Factors

The MULTIPLES of a number are just its <u>times table</u>.

EXAMPLE **Find the first 8 multiples of 13.**

You just need to find the first 8 numbers in the 13 times table:

13   26   39   52   65   78   91   104

The FACTORS of a number are all the numbers that <u>divide into it</u>.

There's a method that guarantees you'll find them all:

1) Start off with 1 × the number itself, then try 2 ×, then 3 × and so on, listing the pairs in rows.

2) Try each one in turn. Cross out the row if it doesn't divide exactly.

3) Eventually, when you get a number <u>repeated</u>, <u>stop</u>.

4) The numbers in the rows you haven't crossed out make up the list of factors.

EXAMPLE **Find all the factors of 24.**

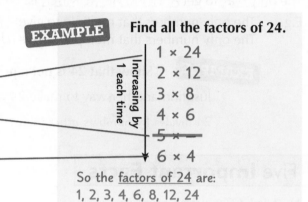

So the <u>factors of 24</u> are:
1, 2, 3, 4, 6, 8, 12, 24

Finding **Prime Factors** — The **Factor Tree**

<u>Any number</u> can be broken down into a string of prime numbers all multiplied together — this is called '<u>expressing it as a product of prime factors</u>', or its '<u>prime factorisation</u>'.

EXAMPLE **Express 420 as a product of prime factors.**

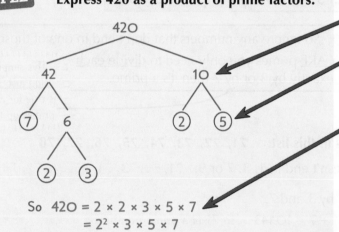

So 420 = 2 × 2 × 3 × 5 × 7
       = 2² × 3 × 5 × 7

To write a number as a product of its prime factors, use the <u>Factor Tree</u> method:

1) Start with the number at the top, and <u>split</u> it into <u>factors</u> as shown.

2) Every time you get a prime, <u>ring it</u>.

3) Keep going until you can't go further (i.e. you're just left with primes), then write the primes out <u>in order</u>. If there's more than one of the <u>same factor</u>, you can write them as <u>powers</u>.

No matter which numbers you choose at each step, you'll find that the prime factorisation is exactly the same. Each number has a <u>unique</u> set of prime factors.

Follow the methods above to find factors and prime factors

Make sure you know the Factor Tree method inside out, then give these Exam Practice Questions a go...

Q1   Use the following list of numbers to answer the questions below. 4, 6, 10, 14, 15, 17, 24, 30
    a) Find one number that's a multiple of 2, a multiple of 3 and a multiple of 4.   [1 mark]
    b) Find one number that's a multiple of 3 and a factor of 36.   [2 marks]

Q2   What number should replace the ☐ to make 14 × 30 = 7 × ☐ true?   [1 mark]

Q3   Express 990 as a product of its prime factors.   [2 marks]

Q3 Video Solution

LCM and HCF

*Two big fancy names but don't be put off — they're both **real easy**. There are two methods for finding each — this page starts you off with the **nice, straightforward** methods.*

LCM — 'Lowest Common Multiple'

'Lowest Common Multiple' — sure, it sounds kind of complicated, but all it means is this:

The SMALLEST number that will DIVIDE BY ALL the numbers in question.

METHOD:
1) LIST the MULTIPLES of ALL the numbers.
2) Find the SMALLEST one that's in ALL the lists.

The LCM is sometimes called the Least (instead of 'Lowest') Common Multiple.

EXAMPLE **Find the lowest common multiple (LCM) of 12 and 15.**

Multiples of 12 are: 12, 24, 36, 48, (60,) 72, 84, 96, ...
Multiples of 15 are: 15, 30, 45, (60,) 75, 90, 105, ...

So the lowest common multiple (LCM) of 12 and 15 is 60.

HCF — 'Highest Common Factor'

'Highest Common Factor' — all it means is this:

The BIGGEST number that will DIVIDE INTO ALL the numbers in question.

METHOD:
1) LIST the FACTORS of ALL the numbers.
2) Find the BIGGEST one that's in ALL the lists.

EXAMPLE **Find the highest common factor (HCF) of 36, 54, and 72.**

Factors of 36 are: 1, 2, 3, 4, 6, 9, 12, (18,) 36
Factors of 54 are: 1, 2, 3, 6, 9, (18,) 27, 54
Factors of 72 are: 1, 2, 3, 4, 6, 8, 9, 12, (18,) 24, 36, 72

So the highest common factor (HCF) of 36, 54 and 72 is 18.

Just take care listing the factors — make sure you use the proper method (as shown on the previous page) or you'll miss one and blow the whole thing out of the water.

LCM and HCF — learn what the names mean

LCM and HCF questions shouldn't be too bad as long as you know exactly what's meant by each of the terms. Now, it's time to make sure you're happy with this stuff by trying out some lovely exam practice questions...

Q1   Find the lowest common multiple (LCM) of 12, 14 and 21.   [2 marks]

Q2   Find the highest common factor (HCF) of 36 and 84.   [2 marks]

LCM and HCF

*The two **methods** on this page are a **little trickier** — but you might have to use them in your exam.*

LCM — Alternative Method

If you already know the <u>prime factors</u> of the numbers, you can use this method instead:

1) List all the <u>PRIME FACTORS</u> that appear in <u>EITHER</u> number.

2) If a factor appears <u>MORE THAN ONCE</u> in one of the numbers, list it <u>THAT MANY TIMES</u>.

3) <u>MULTIPLY</u> these together to give the <u>LCM</u>.

EXAMPLE **$18 = 2 \times 3^2$ and $30 = 2 \times 3 \times 5$. Find the LCM of 18 and 30.**

$18 = 2 \times 3 \times 3$ $30 = 2 \times 3 \times 5$

So the prime factors that appear in either number are: 2, 3, 3, 5

$LCM = 2 \times 3 \times 3 \times 5 = 90$

List 3 twice as it appears twice in 18.

HCF — Alternative Method

Again, there's a different method you can use if you already know the <u>prime factors</u> of the numbers:

1) List all the <u>PRIME FACTORS</u> that appear in <u>BOTH</u> numbers.

2) <u>MULTIPLY</u> these together to find the HCF.

EXAMPLE **$180 = 2^2 \times 3^2 \times 5$ and $84 = 2^2 \times 3 \times 7$. Use this to find the HCF of 180 and 84.**

$180 = ②\times②\times③\times 3 \times 5$ $84 = ②\times②\times③\times 7$

2, 2 and 3 are prime factors of both numbers, so

$HCF = 2 \times 2 \times 3 = 12$

Real-Life LCM and HCF Questions

You might be asked a wordy real-life LCM or HCF question in your exam — these can be <u>tricky</u> to spot at first, but once you have done, the method's <u>just the same</u>.

 Maggie is making party bags. She has 60 balloons, 48 lollipops and 84 stickers. She wants to use them all. Each type of item must be distributed equally between the party bags. What is the maximum number of party bags she can make?

Factors of 60 are: 1, 2, 3, 4, 5, 6, 10, (12,) 15, 20, 30, 60

Factors of 48 are: 1, 2, 3, 4, 6, 8, (12,) 16, 24, 48

Factors of 84 are: 1, 2, 3, 4, 6, 7, (12,) 14, 21, 28, 42, 84

The <u>highest common factor</u> (HCF) of 60, 48 and 84 is 12, so the maximum number of party bags Maggie can make is 12.

You could use the <u>prime factorisation</u> method here if you wanted — use whichever method's <u>easier</u> for you.

In each bag there will be $60 \div 12 = 5$ balloons, $48 \div 12 = 4$ lollipops and $84 \div 12 = 7$ stickers.

You might need to use these alternative methods in the exam

Have a go at these Exam Practice Questions to see if you've got the hang of this LCM and HCF business.

Q1 $112 = 2^4 \times 7$ and $140 = 2^2 \times 5 \times 7$. Use this to find the HCF of 112 and 140. [1 mark]

Q2   A café sells different cakes each day. It sells banana cake every 8 days, and raspberry cake every 22 days. If the café is selling the two cakes today, how many days will it be until the cakes are both sold on the same day again? [3 marks]

Q2 Video Solution

Warm-Up and Worked Exam Questions

These warm-up questions will test whether you've learned the facts from the last few pages.
Keep practising any you get stuck on, before moving on.

Warm-up Questions

1) Which of the following numbers are prime? 30, 31, 32, 33, 34, 35, 36, 37, 38, 39, 40.
2) Explain why 27 is not a prime number.
3) Find all the factors of 40.
4) Find the prime factors of 40.
5) Find the lowest common multiple of 4 and 5.
6) Find the highest common factor of 32 and 88.

Worked Exam Questions

Time for more exam questions with the answers filled in. Understanding these solutions will help
you with the exam questions that follow and in the exam itself.

1   Write down: (3)

   Think of pairs of factors, for example 1 and 28, 2 and 14....

  a)   all the factors of 28,

                                                        1, 2, 4, 7, 14, 28

                                                        [2 marks]

  b)   all the multiples of 8 which appear in the list below.

        55 56 57 58 59 60 61 62 63 64 65

                                                    56, 64

                                                        [1 mark]

2   Dev is making jam. (5)

He needs to buy mini jam jars which come in packs of 12, lids which come in packs of 16 and labels which come in packs of 36. He doesn't want to have any items left over.

Find the smallest number of packs of each item he can buy.

Multiples of 12 are: 12, 24, 36, 48, 60, 72, 84, 96, 108, 120, 132, <u>144</u>, 156, ...

Multiples of 16 are: 16, 32, 48, 64, 80, 96, 112, 128, <u>144</u>, 160, ...

Multiples of 36 are: 36, 72, 108, <u>144</u>, 180, ...

The LCM of 12, 16 and 36 is 144, which is the minimum number of each item he needs.

The minimum number of packs of jars he needs is 144 ÷ 12 = 12 packs

The minimum number of packs of lids he needs is 144 ÷ 16 = 9 packs

The minimum number of packs of labels he needs is 144 ÷ 36 = 4 packs

        12 packs of jars,   **9** packs of lids and   **4** packs of labels

                                                                             [3 marks]

Exam Questions

3   Jasmine says, "there are no prime numbers between 100 and 110."
    Is she correct? Give evidence for your answer.   ③ GRADE

[2 marks]

4   Mei thinks of a prime number. The sum of its digits is one more than a square number.   ③ GRADE
    Write down one number Mei could be thinking of.

....
[2 marks]

5   Write 72 as a product of its prime factors.   ④ GRADE

Make sure your answer only uses prime numbers. Multiply them all together to check you get the number you started with.

....
[2 marks]

6   $P = 3^7 \times 11^2$ and $Q = 3^4 \times 7^3 \times 11$.   ⑤ GRADE

    Write as the product of prime factors:
    a)   the LCM of P and Q,

....
[1 mark]

    b)   the HCF of P and Q.

....
[1 mark]

Fractions without a Calculator

*These pages show you how to cope with fraction calculations without your **beloved calculator**.*

1) Cancelling down

To <u>cancel down</u> or <u>simplify</u> a fraction, <u>divide top and bottom by the same number</u>, till they won't go further:

EXAMPLE Simplify $\frac{18}{24}$.

Cancel down in a series of <u>easy steps</u> — keep going till the top and bottom don't have <u>any</u> common factors.

$$\frac{18}{24} = \frac{6}{8} = \frac{3}{4}$$
$\div3 \quad \div2$
$\div3 \quad \div2$

> The number on the top of the fraction is the <u>numerator</u>, and the number on the bottom is the <u>denominator</u>.

2) Mixed numbers

<u>Mixed numbers</u> are things like $3\frac{1}{3}$, with an integer part and a fraction part.

<u>Improper fractions</u> are ones where the top number is larger than the bottom number.

You need to be able to convert between the two.

EXAMPLES 1. **Write $4\frac{2}{3}$ as an improper fraction.**

1) Think of the <u>mixed number</u> as an <u>addition</u>:

$4\frac{2}{3} = 4 + \frac{2}{3}$

2) Turn the <u>integer part</u> into a <u>fraction</u>:

$4 + \frac{2}{3} = \frac{12}{3} + \frac{2}{3} = \frac{12 + 2}{3} = \frac{14}{3}$

2. **Write $\frac{31}{4}$ as a mixed number.**

<u>Divide</u> the top number by the bottom.

1) The <u>answer</u> gives the <u>whole number part</u>.

2) The <u>remainder</u> goes <u>on top</u> of the fraction.

$31 \div 4 = 7$ remainder 3 so $\frac{31}{4} = 7\frac{3}{4}$

3) Multiplying

Multiply top and bottom <u>separately</u>. Then <u>simplify</u> your fraction as far as possible.

EXAMPLE Find $\frac{8}{5} \times \frac{7}{12}$.

Multiply the top and bottom <u>separately</u>:

$\frac{8}{5} \times \frac{7}{12} = \frac{8 \times 7}{5 \times 12}$

Then <u>simplify</u> — top and bottom both <u>divide by 4</u>.

$= \frac{56}{60} = \frac{14}{15}$

You have to know how to handle mixed numbers

Mixed numbers look difficult, but they're OK once you've converted them into normal fractions.
If you keep on practising working with fractions, you'll bag some easy marks in the exam.

Fractions without a Calculator

Here are some more tricks for dealing with fractions.

4) Dividing

Turn the 2nd fraction <u>UPSIDE DOWN</u> and then <u>multiply</u>:

> When you're multiplying or dividing with <u>mixed numbers</u>, <u>always</u> turn them into improper fractions first.

EXAMPLE   Find $2\frac{1}{3} \div 3\frac{1}{2}$.

Rewrite the <u>mixed numbers</u> as improper <u>fractions</u>: $2\frac{1}{3} \div 3\frac{1}{2} = \frac{7}{3} \div \frac{7}{2}$

Turn $\frac{7}{2}$ <u>upside down</u> and <u>multiply</u>: $= \frac{7}{3} \times \frac{2}{7} = \frac{7 \times 2}{3 \times 7}$

<u>Simplify</u> — top and bottom both <u>divide by 7</u>. $= \frac{14}{21} = \frac{2}{3}$

5) Common denominators

This comes in handy for <u>ordering fractions</u> by size, and for <u>adding</u> or <u>subtracting</u> fractions.

You need to find a number that <u>all</u> the denominators <u>divide into</u> — this will be your <u>common denominator</u>. The simplest way is to find the <u>lowest common multiple</u> of the denominators:

EXAMPLE

Put these fractions in ascending order of size:

$$\frac{8}{3} \qquad \frac{5}{4} \qquad \frac{12}{5}$$

The <u>LCM</u> of 3, 4 and 5 is 60, so make 60 the <u>common denominator</u>:

$$\frac{8}{3} \overset{\times 20}{=} \frac{160}{60} \qquad \frac{5}{4} \overset{\times 15}{=} \frac{75}{60} \qquad \frac{12}{5} \overset{\times 12}{=} \frac{144}{60}$$

So the correct order is $\frac{75}{60}, \frac{144}{60}, \frac{160}{60}$ i.e. $\frac{5}{4}, \frac{12}{5}, \frac{8}{3}$

> Don't forget to use the original fractions in the final answer.

Common denominators are really handy

You need to find a common denominator to order, add or subtract fractions (see next page). If you're ordering fractions, don't forget to turn them back into their original forms when you give your answers.

Fractions without a Calculator

6) **Adding**, **subtracting** — sort the denominators first

If you're adding or subtracting <u>mixed numbers</u>, it usually helps to convert them to improper fractions first.

> Make sure the denominators are <u>the same</u> (see previous page).
> Add (or subtract) the top lines <u>only</u>.

EXAMPLE Calculate $2\frac{1}{5} - 1\frac{1}{2}$.

Rewrite the <u>mixed numbers</u> as improper <u>fractions</u>: $2\frac{1}{5} - 1\frac{1}{2} = \frac{11}{5} - \frac{3}{2}$

Find a <u>common denominator</u>: $= \frac{22}{10} - \frac{15}{10}$

Combine the <u>top lines</u>: $= \frac{22-15}{10} = \frac{7}{10}$

7) **Fractions of something**

EXAMPLE What is $\frac{9}{20}$ of £360?

'$\frac{9}{20}$ of £360' means '$\frac{9}{20} \times £360$'.

<u>Multiply</u> by the top of the fraction and <u>divide</u> by the bottom.

$\frac{9}{20} \times £360 = (£360 \div 20) \times 9$
$= £18 \times 9 = £162$

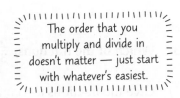
The order that you multiply and divide in doesn't matter — just start with whatever's easiest.

8) **Expressing** as a **Fraction**

EXAMPLE Write 180 as a fraction of 80.

Write the first number over the second and <u>cancel down</u>. $\frac{180}{80} = \frac{9}{4}$

You have to learn how to handle fractions in these 8 situations

When you think you've got the hang of it all, try these Exam Practice Questions without a calculator.

Q1 Calculate: a) $\frac{5}{8} \times 1\frac{5}{6}$ [3 marks] b) $\frac{10}{7} \div \frac{8}{3}$ [2 marks]

c) $\frac{8}{9} + \frac{19}{27}$ [2 marks] d) $5\frac{2}{3} - 9\frac{1}{4}$ [3 marks]

Q2 Dean has made 30 sandwiches. $\frac{7}{15}$ of the sandwiches he has made are vegetarian, and $\frac{3}{7}$ of the vegetarian sandwiches are cheese sandwiches. How many cheese sandwiches has he made? [2 marks] Q2 Video Solution

Fraction Problems

*The previous three pages gave you all the tools you'll need to tackle these more **pesky fraction questions**. All these questions could come up on the **non-calculator** paper so put your calculators away.*

Use the **Methods** You've **Already Learnt**

 EXAMPLES

1. **Rachel is making cupcakes. Each cupcake needs $\frac{2}{25}$ of a pack of butter. How many packs of butter will Rachel need to buy to make 30 cupcakes?**

Once you've done the <u>multiplication</u>, convert $\frac{60}{25}$ to a <u>mixed number</u> to see how many packs she'll need — careful though, she can only buy a <u>whole number</u> of packs.

1 cupcake needs $\frac{2}{25}$ of a pack of butter,

so 30 cupcakes need: $30 \times \frac{2}{25} = \frac{60}{25} = \frac{12}{5} = 2\frac{2}{5}$ packs

2 packs won't be enough, so Rachel will need to buy 3 packs.

2. **The diamond to the right is made up of two identical equilateral triangles. The equilateral triangle at the top is split into three equal triangles. The equilateral triangle at the bottom is split into two equal triangles. Find the fraction of the diamond that is shaded.**

$\frac{2}{3}$ of the top triangle is shaded. This is $\frac{2}{3} \times \frac{1}{2} = \frac{2}{6} = \frac{1}{3}$ of the diamond.

$\frac{1}{2}$ of the bottom triangle is shaded. This is $\frac{1}{2} \times \frac{1}{2} = \frac{1}{4}$ of the diamond.

Remember to <u>multiply by a half</u> because each equilateral triangle makes up <u>half the diamond</u>.

Find a <u>common denominator</u> to add the fractions. $\frac{1}{3} + \frac{1}{4} = \frac{4}{12} + \frac{3}{12} = \frac{7}{12}$

3. **Which of the fractions, $\frac{2}{3}$ or $\frac{9}{7}$, is closer to 1?**

1) Give the fractions a <u>common denominator</u>. $\quad \frac{2}{3} = \frac{2 \times 7}{3 \times 7} = \frac{14}{21} \qquad \frac{9}{7} = \frac{9 \times 3}{7 \times 3} = \frac{27}{21}$

2) Convert 1 into a fraction and <u>subtract</u> to see which fraction is <u>closer</u> to 1. $\quad \frac{21}{21} - \frac{14}{21} = \frac{7}{21} \qquad \frac{27}{21} - \frac{21}{21} = \frac{6}{21} \qquad$ So $\frac{9}{7}$ is closer to 1.

4. **At the Prism School, year 10 is split into two classes, each with the same number of pupils in total. $\frac{3}{5}$ of one class are girls, and $\frac{4}{7}$ of the other class are girls. What fraction of year 10 students are girls?**

1) <u>Divide</u> each fraction by <u>2</u> to find the number of girls in each class as a fraction of <u>total year 10 pupils</u>.

Class 1 girls are: $\frac{3}{5} \div 2 = \frac{3}{10}$ of the total pupils in year 10.

Class 2 girls are: $\frac{4}{7} \div 2 = \frac{4}{14}$ of the total pupils in year 10.

2) Find a <u>common denominator</u>, then <u>add</u> the fractions.

So $\frac{3}{10} + \frac{4}{14} = \frac{21}{70} + \frac{20}{70} = \frac{41}{70}$ of year 10 pupils are girls.

Practise using the different fraction rules

A lot of questions require you to find a common denominator, so make sure you're comfortable with that.

Q1 Bill has some books. They're all fiction books, biographies or history books. $\frac{3}{14}$ of the books are biographies and $\frac{1}{6}$ are history books. What fraction of Bill's books are fiction books? Give your answer as a fraction in its simplest form. [4 marks]

Q1 Video Solution

Fractions, Decimals and Percentages

Fractions, decimals and percentages are **three different ways** of describing when you've got **part** of a **whole thing**. They're **closely related** and you can **convert between them**. These tables show some really common conversions which you should know straight off without having to work them out:

Fraction	Decimal	Percentage
$\frac{1}{2}$	0.5	50%
$\frac{1}{4}$	0.25	25%
$\frac{3}{4}$	0.75	75%
$\frac{1}{3}$	0.333333...	$33\frac{1}{3}$%
$\frac{2}{3}$	0.666666...	$66\frac{2}{3}$%
$\frac{5}{2}$	2.5	250%

Fraction	Decimal	Percentage
$\frac{1}{10}$	0.1	10%
$\frac{2}{10}$	0.2	20%
$\frac{1}{5}$	0.2	20%
$\frac{2}{5}$	0.4	40%
$\frac{1}{8}$	0.125	12.5%
$\frac{3}{8}$	0.375	37.5%

The more of those conversions you learn, the better — but for those that you <u>don't know</u>, you must <u>also learn</u> how to <u>convert</u> between the three types. These are the methods:

$$\text{Fraction} \xrightarrow{\text{Divide}} \text{Decimal} \xrightarrow{\times \text{ by } 100} \text{Percentage}$$

E.g. $\frac{7}{20}$ is $7 \div 20$ $= 0.35$ e.g. 0.35×100 $= 35\%$

$$\text{Fraction} \xleftarrow{\text{The awkward one}} \text{Decimal} \xleftarrow{\div \text{ by } 100} \text{Percentage}$$

<u>Converting decimals to fractions</u> is awkward. To convert terminating decimals to fractions:

The digits after the decimal point go on the top, and a <u>10, 100, 1000, etc.</u> on the bottom — so you have the same number of zeros as there were decimal places.

$0.6 = \frac{6}{10}$ $0.78 = \frac{78}{100}$ $0.024 = \frac{24}{1000}$ etc.   These can often be <u>cancelled down</u> — see p.15.

Recurring Decimals have Repeating Digits

1) Recurring decimals have a <u>pattern of numbers</u> which <u>repeats forever</u>, e.g. 0.333333... which is $\frac{1}{3}$.
2) The <u>repeating part</u> is usually marked with <u>dots</u> on top of the number.
3) If there's <u>one dot</u>, only <u>one digit</u> is repeated. If there are <u>two dots</u>, <u>everything from the first dot to the second dot</u> is the repeating bit.
4) You can <u>convert</u> a fraction to a recurring decimal:

E.g. $0.2\dot{5} = 0.2555555...$,
$0.\dot{2}\dot{5} = 0.25252525...$,
$0.\dot{2}6\dot{5} = 0.265265265...$

EXAMPLE   Write $\frac{5}{11}$ as a recurring decimal.

Just do the division, and look for the <u>repeating pattern</u>. $5 \div 11 = 0.454545...$ so $\frac{5}{11} = 0.\dot{4}\dot{5}$

Fractions, decimals and percentages are interchangeable

You just need to remember how to switch between them. Give it a go with these Exam Practice Questions.

Q1   Which is greater: a) 57% or $\frac{5}{9}$,   b) 0.2 or $\frac{6}{25}$,   c) $\frac{7}{8}$ or 90%?   [3 marks]

Q2   a) Write 0.555 as a fraction in its simplest form.   [2 marks]

       b) Write $\frac{1}{6}$ as a recurring decimal.   [2 marks]

Warm-up and Worked Exam Questions

Here's a set of warm-up questions for this section. Work through them to check you've got the hang of fractions and to limber up for the exam questions that follow.

Warm-up Questions

1) Simplify $\frac{48}{64}$ as far as possible.

2) Which of these fractions are equivalent to $\frac{1}{3}$? $\frac{2}{6}, \frac{5}{15}, \frac{9}{36}, \frac{6}{20}$

3) Work these out, then simplify your answers where possible:

   a) $\frac{2}{5} \times \frac{2}{3}$ b) $\frac{2}{5} \div \frac{2}{3}$ c) $\frac{2}{5} + \frac{2}{3}$ d) $\frac{2}{3} - \frac{2}{5}$

4) What decimal is the same as $\frac{7}{10}$?

5) What percentage is the same as $\frac{2}{3}$?

6) What fraction is the same as 0.4?

7) Write $\frac{2}{7}$ as a recurring decimal.

Worked Exam Questions

Make sure you understand what's going on in these questions before trying the next page for yourself.

1   Sarah and her 4 friends eat $\frac{5}{6}$ of a pizza each. Pizzas cost £4.50 each, or 2 for £7. **(4)**

What is the minimum amount they will have to spend on pizzas?

$$5 \times \frac{5}{6} = \frac{25}{6}$$

$$= 4\frac{1}{6}, \text{ so they will need 5 pizzas in total.}$$

$$\text{Cost} = \underline{(2 \times £7)} + £4.50$$

$$= £14 + £4.50$$

They can get 4 pizzas in the 2 for £7 offer.

$$= £18.50$$

£ ......18.50......

[3 marks]

2   *ABC* is an equilateral triangle. It has been divided into smaller equilateral triangles as shown below. **(4)**

What fraction of triangle *ABC* is shaded?

Find the shaded areas as separate fractions of the whole triangle, then add them up.

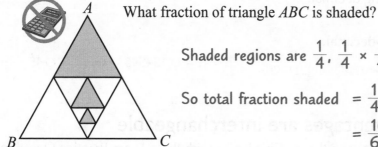

Shaded regions are $\frac{1}{4}$, $\frac{1}{4} \times \frac{1}{4} = \frac{1}{16}$ and $\frac{1}{4} \times \frac{1}{4} \times \frac{1}{4} = \frac{1}{64}$

So total fraction shaded $= \frac{1}{4} + \frac{1}{16} + \frac{1}{64}$

$$= \frac{16}{64} + \frac{4}{64} + \frac{1}{64}$$

$$= \frac{21}{64}$$

$\frac{21}{64}$

....

[3 marks]

Exam Questions

3   The number of people at last Saturday's Norchester City game was 12 400.

Season ticket holders made up $\frac{3}{8}$ of the crowd. How many season ticket holders were there?

....

[2 marks]

4   Write the following in order of size, starting with the smallest.

Start by writing all the
numbers as decimals.   65% $\frac{2}{3}$ 0.065 $\frac{33}{50}$

.... , ... , ... , ...

[2 marks]

5   Work out the following, giving your answers as mixed numbers in their simplest form:

 a) $1\frac{1}{8} \times 2\frac{2}{5}$

....

[3 marks]

b) $1\frac{3}{4} \div \frac{7}{9}$

....

[3 marks]

6   Bilal, Eli, Ruth and Jenny split the bill at a restaurant. Bilal pays $\frac{1}{4}$ of the bill
and Eli and Ruth each pay 20% of the bill. Jenny pays £17.50.

   How much was the bill in total?

£ ....

[4 marks]

Rounding Numbers

*You need to be able to use **3 different rounding methods**.*
We'll do decimal places first, but there's the same basic idea behind all three.

Decimal Places (d.p.)

To round to a given number of <u>decimal places</u>:

> 1) <u>Identify</u> the position of the '<u>last digit</u>' from the number of decimal places.
> 2) Then look at the next digit to the <u>right</u> — called <u>the decider</u>.
> 3) If the <u>decider</u> is <u>5 or more</u>, then <u>round up</u> the <u>last digit</u>.
>   If the <u>decider</u> is <u>4 or less</u>, then leave the <u>last digit</u> as it is.
> 4) There must be <u>no more digits</u> after the last digit (not even zeros).

If you're rounding to 2 d.p. the last digit is the <u>second</u> digit after the decimal point.

EXAMPLES

1. What is 13.72 correct to <u>1 decimal place</u>?

$$13.72 = 13.7$$

<u>LAST DIGIT</u> to be written (1st decimal place because we're rounding to 1 d.p.)

<u>DECIDER</u>

The <u>LAST DIGIT</u> stays the <u>same</u> because the <u>DECIDER</u> is <u>4 or less</u>.

2. What is 7.45839 to <u>2 decimal places</u>?

$$7.45839 = 7.46$$

<u>LAST DIGIT</u> to be written (2nd decimal place because we're rounding to 2 d.p.)

<u>DECIDER</u>

The <u>LAST DIGIT</u> rounds <u>UP</u> because the <u>DECIDER</u> is <u>5 or more</u>.

Watch Out for **Pesky Nines**

If you have to <u>round up</u> a <u>9</u> (to 10), replace the 9 with 0, and <u>add 1</u> to the digit on the <u>left</u>.

decider

E.g. Round 45.698 to 2 d.p: $45.698 \longrightarrow 45.69 \longrightarrow 45.70$ to 2 d.p.

last digit — round up

The question asks for 2 d.p. so you <u>must</u> put 45.<u>70</u> not 45.7.

You might be asked to round off your answers in the exam

This is important stuff, so learn the steps of the basic method and then have a go at these:

Q1 a) Give 21.435 correct to 1 decimal place
    b) Give 0.0581 correct to 2 d.p.
    c) Round 4.968 to 1 d.p.   [3 marks]

Q2 Calculate $\frac{25.49 - 16.73}{2.82}$
    and give your answer to 2 d.p. [2 marks]

Rounding Numbers

Significant Figures (s.f.)

The 1st significant figure of any number is the first digit which isn't a zero.

The 2nd, 3rd, 4th, etc. significant figures follow immediately after the 1st — they're allowed to be zeros.

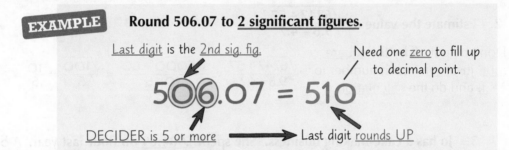

0.002309   **506.07**

SIG. FIGS: 1st 2nd 3rd 4th   1st 2nd 3rd 4th

To round to a given number of significant figures:

1) Find the last digit — if you're rounding to, say 3 s.f., then the 3rd significant figure is the last digit.
2) Use the digit to the right of it as the decider, just like for d.p.
3) Once you've rounded, fill up with zeros, up to but not beyond the decimal point.

EXAMPLE   **Round 506.07 to 2 significant figures.**

Last digit is the 2nd sig. fig.

Need one zero to fill up to decimal point.

5<u>0</u>6.07 = 510

DECIDER is 5 or more ———→ Last digit rounds UP

To the Nearest Whole Number, Ten, Hundred etc.

You might be asked to round to the nearest whole number, ten, hundred, thousand, or million:

1) Identify the last digit, e.g. for the nearest whole number it's the units position, and for the 'nearest ten' it's the tens position, etc.
2) Round the last digit and fill in with zeros up to the decimal point, just like for significant figures.

EXAMPLE   **Round 6751 to the nearest hundred.**

Last digit is in the 'hundreds' position

Fill in 2 zeros up to decimal point.

6<u>7</u>51 = 6800

DECIDER is 5 or more ———→ Last digit rounds UP.

Significant figures can be a bit tricky to get your head round

Learn the method on this page for identifying them, and make sure you really understand the example.

Q1 a) Round 653 to 1 s.f.   b) Round 14.6 to 2 s.f.   Q2 Calculate $\frac{8.43 + 12.72}{5.63 - 1.21}$.
    c) Give 168.7 to the nearest whole number.   Give your answer to 2 s.f.
    d) Give 82 430 to the nearest thousand. [4 marks]   [2 marks]

Q2 Video Solution

Estimating

'Estimate' doesn't mean 'take a wild guess', it means 'look at the numbers, make them a bit easier, then do the calculation'. Your answer won't be as **accurate** as the real thing but it's easier on your brain.

Estimating **Calculations**

1) <u>Round everything off</u> to <u>1 significant figure.</u>
2) Then <u>work out the answer</u> using these nice easy numbers.
3) <u>Show all your working</u> or you won't get the marks.

> Have a look at the previous page to remind yourself how to round to 1 s.f.

EXAMPLES

1. **Estimate the value of 42.6 × 12.1.**

> ≈ means '<u>approximately equal to</u>'.

1) <u>Round</u> each number to <u>1 s.f.</u>   42.6 × 12.1 ≈ 40 × 10
2) Do the <u>calculation</u> with   = 400
   the rounded numbers.

> You might have to say if it's an <u>underestimate</u> or an <u>overestimate</u>. Here, you rounded both numbers <u>down</u>, so it's an <u>underestimate</u>.

2. **Estimate the value of** $\dfrac{\sqrt{6242 \div 57}}{9.8 - 4.7}$.

Don't be put off by the <u>square root</u>, just <u>round</u> each number to <u>1 s.f.</u> and do the <u>calculation</u>.

$$\frac{\sqrt{6242 \div 57}}{9.8 - 4.7} \approx \frac{\sqrt{6000 \div 60}}{10 - 5} = \frac{\sqrt{100}}{5} = \frac{10}{5} = 2$$

3. **Jo has a cake-making business. She spent £984.69 on flour last year. A bag of flour costs £1.89, and she makes an average of 5 cakes from each bag of flour. Work out an estimate of how many cakes she made last year.**

> Don't panic if you get a '<u>real-life</u>' estimating question — just round everything to 1 s.f. as before.

1) Estimate number of bags of flour — <u>round</u> numbers to <u>1 s.f.</u>

Number of bags of flour = $\dfrac{984.69}{1.89}$

$\approx \dfrac{1000}{2} = 500$

2) Multiply to find the number of cakes.

Number of cakes ≈ 500 × 5 = 2500

You Might Need to Estimate **Height**

> Use <u>1.8 m</u> as an estimate for the <u>height of a man</u>.

EXAMPLE   **Estimate the height of the giraffe in the picture.**

In the picture the giraffe's about <u>two and a half times</u> as tall as the man.

Height of a man is about 1.8 m
Rough height of giraffe = 2.5 × height of man
                        = 2.5 × 1.8 = 4.5 m

Round the numbers first, then do the calculation like normal

If you're asked to estimate something in the exam, make sure you show all your steps (including what each number is rounded to) to prove that you didn't just use a calculator. Hassle, but it'll pay off.

Q1   Estimate the value of: a) $\dfrac{22.3 \times 11.4}{0.532}$ [2 marks]   b) $\sqrt{\dfrac{18.79 \times 5.22}{1.23 + 3.21}}$ [2 marks]

Q2   Padma buys 17 kg of turnips at a cost of £1.93 per kg. Estimate the total cost. [2 marks]

Rounding Errors

*Whenever a number is **rounded** or **truncated** it will have some amount of **error**.*
*The error tells you how far the **actual value** could be away from the **rounded value**.*

Rounded Measurements Can Be Out By Half A Unit

> Whenever a measurement is <u>rounded off</u> to a <u>given UNIT</u> the
> <u>actual measurement</u> can be anything up to <u>HALF A UNIT bigger or smaller</u>.

If you're asked for the <u>error interval</u>, you can use <u>inequalities</u>
to find the <u>range of values</u> that the <u>actual measurement</u> could be:

EXAMPLE   **The mass of a cake is given as 2.4 kg to the nearest 0.1 kg.**
Find the interval within which m, the actual mass of the cake, lies.

It's measured to the nearest 0.1 kg, so the actual mass could be 0.1 ÷ 2 = 0.05 kg bigger or smaller.

Minimum mass = 2.4 − 0.05 = <u>2.35</u> kg
Maximum mass = 2.4 + 0.05 = <u>2.45</u> kg

So the interval is 2.35 kg ≤ m < 2.45 kg

The actual maximum mass is 2.44999... kg, but it's OK to say 2.45 kg instead.

The actual value is <u>greater than or equal to</u> the <u>minimum</u> but <u>strictly less than</u> the <u>maximum</u>.
The actual mass of the cake could be <u>exactly</u> 2.35 kg,
but if it was exactly 2.45 kg it would <u>round up</u> to 2.5 kg instead.

If a question involves <u>more than one</u> rounded value,
you need to think about the range of <u>possible values</u> for <u>each one</u>.

EXAMPLE   **The length of rope A is 120 cm to the nearest 10 cm.**
The length of rope B is 200 cm to the nearest 50 cm.
Work out the maximum total length of the two ropes.

Find the <u>maximum</u>
length of <u>each</u> rope.

Maximum length of rope A = 120 + 5 = 125 cm
Maximum length of rope B = 200 + 25 = 225 cm

So the maximum total length is 125 cm + 225 cm = 350 cm

Truncated Measurements Can Be A Whole Unit Out

You truncate a number by <u>chopping off</u> decimal places.   E.g. 25.765674 truncated to 1 d.p. would be 25.7

> When a measurement is <u>TRUNCATED</u> to a <u>given UNIT</u>, the
> <u>actual measurement</u> can be up to <u>A WHOLE UNIT bigger but no smaller</u>.

If the mass of the cake in the first example above was 2.4 kg <u>truncated</u> to 1 d.p.,
the error interval would be 2.4 kg ≤ m < 2.5 kg.
So even if the mass was 2.499999 kg, it would still truncate to 2.4 kg.

Learn the difference between rounding and truncating

You need to be comfortable with rounding before you tackle this topic — so if you're feeling a bit unsure,
take a look back at those pages. Feeling confident? Then have a go at these questions now.

Q1   The following numbers have been rounded to 2 significant figures.
   Give the error interval for each: a) 380   [2 marks]   b) 0.46   [2 marks]

Q2   A bag of apples weighs 800 g to the nearest 100 g.
   A bag of potatoes weighs 1.5 kg to the nearest 500 g.
   What is the minimum possible total weight of the apples and potatoes?   [3 marks]

Q2 Video Solution

Warm-up and Worked Exam Questions

Before you dive into the exam questions on the next page, have a paddle in these friendly-looking warm-up questions. If you're not sure about any of them, go back and look at the topic again.

Warm-up Questions

1) Round these numbers off to 1 decimal place:
   a) 3.24   b) 1.78   c) 2.31   d) 0.46   e) 9.76

2) Round these off to the nearest whole number:
   a) 3.4   b) 5.2   c) 1.84   d) 6.9   e) 3.26

3) Round these numbers to the stated number of significant figures:
   a) 352 to 2 s.f.   b) 465 to 1 s.f.   c) 12.38 to 3 s.f.   d) 0.03567 to 2 s.f.

4) Round these numbers off to the nearest hundred:
   a) 2865   b) 450   c) 123

5) a) Estimate (29.5 – 9.6) × 4.87   b) Is your answer an underestimate or an overestimate?

6) A bottle contains 568 ml of milk to the nearest ml.
   What is the minimum possible volume of milk in the bottle?

7) The following numbers have been rounded as shown.
   Give the error interval for each: a) 0.915 (to 3 d.p.)   b) 1120 (to 3 s.f.)

8) Truncate: a) 37.919 to 1 d.p.   b) 2.1503 to 1 d.p.

Worked Exam Question

Look at that — an exam question with all the answers filled in. How unexpected.

1   Round the following to the given degree of accuracy. (2)

a)   Kudzai has 123 people coming to his party.
      Write this number to the nearest 10.

      1(2)3 ——————— Last digit is in the 'tens' position
      12(3) ——————— Decider is less than 5
      120 ——————— 1 space to fill before the decimal point   120
                                                                    [1 mark]

b)   The attendance at a football match was 2568 people.
      What is this to the nearest hundred?

      Last digit   Decider
      2(5)6 8
      2600 ——————— 2 spaces to fill   2600
                                                                    [1 mark]

c)   The population of Ulverpool is 452 529.
      Round this to the nearest 100 000.

      Last digit   Decider
      (4)52 529
      500 000 ——————— 5 spaces to fill   500 000
                                                                    [1 mark]

Exam Questions

2   The distance between two stars is 428.6237 light years. ③

    a)   Round this distance to one decimal place.

.... light years
[1 mark]

    b)   Round this distance to 2 significant figures.

.... light years
[1 mark]

3   A stall sells paperback and hardback books. Paperback books cost £4.95 and hardback books cost £11. One Saturday, the stall sells 28 paperback and 19 hardback books. ④

    a)   Find an estimate for the amount of money the stall made that day.
        Show all your working.

£ ....
[2 marks]

    b)   The actual amount the stall made was £347.60.
        Do you think your estimate was sensible? Explain your answer.

....

....
[1 mark]

4   Estimate the value of $\dfrac{12.2 \times 1.86}{0.19}$ ④

You should start by rounding each number to an easier one.

....
[2 marks]

5   Joseph is weighing himself. His scales give his weight to the nearest kilogram. ⑤

According to his scales, Joseph is 57 kg.
What are the minimum and maximum weights that he could be?

Minimum weight: .... kg

Maximum weight: .... kg
[2 marks]

6   Given that $a = 3.8$ to 1 decimal place, ⑤
write down the error interval for a.

....
[2 marks]

Powers

You've already seen 'to the power 2' and 'to the power 3' — they're just 'squared' and 'cubed'.
They're just the tip of the iceberg — any number can be a power if it puts its mind to it...

Powers are a very Useful Shorthand

1) Powers are 'numbers <u>multiplied by themselves</u> so many times':

$$2 \times 2 \times 2 \times 2 \times 2 \times 2 \times 2 = 2^7 \quad \text{('two to the power 7')}$$

2) The <u>powers of ten</u> are really easy — the power tells you the number of zeros:

   to the power of 6

   $$10^1 = 10 \qquad 10^2 = 100 \qquad 10^3 = 1000 \qquad 10^6 = 1\,000\,000$$

   6 zeros

3) Use the $x^\blacksquare$ or y^x button on your calculator to find powers,
   e.g. press 3 · 7 $x^\blacksquare$ 3 = to get $3.7^3 = 50.653$.

4) Anything to the <u>power 1</u> is just <u>itself</u>, e.g. $4^1 = 4$.

5) <u>1 to any power</u> is <u>still 1</u>, e.g. $1^{457} = 1$.

6) <u>Anything</u> to the <u>power 0</u> is just <u>1</u>, e.g. $5^0 = 1$, $67^0 = 1$, $x^0 = 1$.

Four Easy Rules:

> **Warning**: Rules 1 & 2 <u>don't work</u> for things like $2^3 \times 3^7$, only for <u>powers of the same number</u>.

1) When <u>MULTIPLYING</u>, you <u>ADD THE POWERS</u>. e.g. $3^4 \times 3^6 = 3^{4+6} = 3^{10}$

2) When <u>DIVIDING</u>, you <u>SUBTRACT THE POWERS</u>. e.g. $c^4 \div c^2 = c^{4-2} = c^2$

3) When <u>RAISING</u> one power to another, you <u>MULTIPLY THE POWERS</u>. e.g. $(3^2)^4 = 3^{2 \times 4} = 3^8$

4) <u>FRACTIONS</u> — Apply the power to <u>both TOP and BOTTOM</u>. e.g. $\left(\frac{2}{3}\right)^3 = \frac{2^3}{3^3} = \frac{8}{27}$

EXAMPLE $a = 5^9$ and $b = 5^4 \times 5^2$. **What is the value of $\frac{a}{b}$?**

1) Work out b — <u>add</u> the powers: $b = 5^4 \times 5^2 = 5^{4+2} = 5^6$

2) <u>Divide</u> a by b — <u>subtract</u> the powers: $\frac{a}{b} = 5^9 \div 5^6 = 5^{9-6} = 5^3 = 125$

One Trickier Rule

People have real difficulty remembering this — whenever you see a <u>negative power</u> you need to immediately think: "Aha, that means turn it the other way up and make the power positive".

To find a <u>negative power</u> — turn it <u>upside-down</u>.

E.g. $7^{-2} = \frac{1}{7^2} = \frac{1}{49}$, $\left(\frac{3}{5}\right)^{-2} = \left(\frac{5}{3}\right)^2 = \frac{5^2}{3^2} = \frac{25}{9}$

Practise these power rules — you never know when you'll need them

If you can add, subtract and multiply, there's nothing on this page you can't do — just learn the rules.

Q1 a) Find $3^3 + 4^2$ without a calculator.
   b) Use your calculator to find 6.2^3.
   c) Write ten thousand as a power of 10.
   [4 marks]

Q2 Simplify: a) $4^2 \times 4^3$ b) $\frac{7^6}{7^3}$ c) $(q^2)^4$ [3 marks]

Q3 Without using a calculator, find
   a) $\frac{6^3 \times 6^5}{6^6}$ b) 2^{-4} [4 marks]

Q3 Video Solution

Roots

Take a deep breath, and get ready to tackle this page.

Square Roots

'Squared' means 'multiplied by itself': $8^2 = 8 \times 8 = 64$

SQUARE ROOT $\sqrt{}$ is the reverse process: $\sqrt{64} = 8$

> **'Square Root' means 'What Number Times by Itself gives...'**

 EXAMPLES

1. What is $\sqrt{49}$?

7 times by itself gives 49: $49 = 7 \times 7$

So $\sqrt{49} = 7$

> 49 is a <u>square number</u> — make sure you know all the <u>square numbers</u> on p.1 so you can answer questions like this <u>without a calculator</u>.

2. What is $\sqrt{29.16}$?

Use your calculator. Press: $\boxed{\sqrt{}}$ $\boxed{29.16}$ $\boxed{=}$ 5.4

3. Find <u>both</u> square roots of 36.

$6 \times 6 = 36$, so positive square root = 6

$-6 \times -6 = 36$, so negative square root = -6

> **All numbers also have a NEGATIVE SQUARE ROOT — it's just the '–' version of the normal positive one.**

This little rule for multiplying roots might come in <u>handy</u> in your exam: $\sqrt{a} \times \sqrt{a} = a$

> <u>Be careful</u> — this is <u>only true</u> if you're multiplying together two roots which are the <u>same</u>.

Cube Roots

'Cubed' means 'multiplied by itself and then by itself again': $2^3 = 2 \times 2 \times 2 = 8$

CUBE ROOT $\sqrt[3]{}$ is the reverse process: $\sqrt[3]{8} = 2$

> **'Cube Root' means 'What Number Times by Itself and then by Itself Again gives...'**

You need to be able to write down the cube roots of the <u>cube numbers</u> given on p.1 <u>without a calculator</u>.

To find the cube root of any other number you can use your calculator — press $\boxed{\sqrt[3]{}}$.

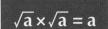 **EXAMPLES**

1. What is $\sqrt[3]{27}$?

> 27 is a <u>cube number</u>.

3 times by itself and then by itself again gives 27: $27 = 3 \times 3 \times 3$

So $\sqrt[3]{27} = 3$

2. What is $\sqrt[3]{4913}$?

Press: $\boxed{\sqrt[3]{}}$ $\boxed{4913}$ $\boxed{=}$ 17

You can use your calculator to find <u>any root</u> of a number, using the $\boxed{\sqrt[x]{}}$ or $\boxed{\sqrt[\square]{\square}}$ buttons.

> Make sure you know how to use your calculator to find higher order roots — the buttons might be slightly different to these ones.

Learn the difference between square roots and square numbers

A square root is a number that, when multiplied by itself, gives the number under the root symbol.

Q1 Find a) $\sqrt{196}$ without using a calculator.   b) $\sqrt[3]{9261}$ c) $\sqrt[7]{2187}$ [3 marks]

Q2 The volume of a cube is 1.728 cm³. Find the length of one of its sides, in cm.   [2 marks]

Q3 Work out $\sqrt[3]{19.34} + (1.3 + 2.5)^2$. Write down the full calculator display.   [1 mark]

Standard Form

*Standard form is useful for writing **VERY BIG** or **VERY SMALL** numbers in a more convenient way.*

E.g. $56\,000\,000\,000$ would be 5.6×10^{10} in standard form.

   $0.000\,000\,003\,45$ would be 3.45×10^{-9} in standard form.

*But **ANY NUMBER** can be written in standard form and you need to know how to do it:*

What it **Actually** is:

A number written in standard form must <u>always</u> be in <u>exactly</u> this form:

This <u>number</u> must <u>always</u> be <u>between 1 and 10</u>.

(The fancy way of saying this is $1 \le A < 10$)

$$A \times 10^n$$

This number is just the <u>number of places</u> the <u>decimal point</u> moves.

Learn the <u>THREE RULES</u>:

1) The <u>front number</u> must always be <u>between 1 and 10</u>.

2) The power of 10, n, is <u>how far the decimal point moves</u>.

3) n is <u>positive for BIG numbers</u>, n is <u>negative for SMALL numbers</u>.

(This is much better than rules based on which way the decimal point moves.)

Four **Important** Examples:

1   Express 35 600 in standard form.

1) <u>Move the decimal point</u> until 35 600 becomes 3.56 ($1 \le A < 10$)

2) The decimal point has moved <u>4 places</u> so n = 4, giving: 10^4 $3\,5\,6\,0\,0\,0.0$

3) 35 600 is a <u>big number</u> so n is +4, not –4 $= 3.56 \times 10^4$

2   Express 0.0000623 in standard form.

1) The decimal point must move <u>5 places</u> to give 6.23 ($1 \le A < 10$). So the power of 10 is 5. $0.0\,0\,0\,0\,6\,2\,3$

2) Since 0.0000623 is a <u>small number</u> it must be 10^{-5} not 10^{+5} $= 6.23 \times 10^{-5}$

3   Express 4.95×10^{-3} as an ordinary number.

1) The power of 10 is <u>negative</u>, so it's a <u>small number</u> — the answer will be less than 1. $0\,0\,0\,4.9\,5 \times 10^{-3}$

2) The power is –3, so the decimal point moves <u>3 places</u>. $= 0.00495$

4   Which is the largest number in the following list? 9.5×10^8 2.7×10^5 3.6×10^8 5.6×10^6

1) Compare the <u>powers</u> first. 9.5×10^8 and 3.6×10^8 have the biggest powers so one of them is the largest.

2) Then, compare the <u>front numbers</u>.   9.5 is greater than 3.6 So 9.5×10^8 is the largest number.

Standard Form

*You might be asked to add, subtract, multiply or divide using numbers in standard form **without** using a calculator.*

Multiplying and Dividing

1) Rearrange to put the <u>front numbers</u> and the <u>powers of 10 together</u>.
2) Multiply or divide the front numbers, and use the <u>power rules</u> (see p.28) to multiply or divide the powers of 10.
3) Make sure your answer is still in <u>standard form</u>.

EXAMPLES

1. **Find $(2 \times 10^3) \times (6 \times 10^5)$ without using a calculator. Give your answer in standard form.**

Multiply front numbers and powers separately ——
$(2 \times 10^3) \times (6 \times 10^5)$
$= (2 \times 6) \times (10^3 \times 10^5)$
$= 12 \times 10^{3+5}$ —— Add the powers (see p.28)

Not in standard form so convert it — divide the number by 10...
$= 12 \times 10^8$
$= 1.2 \times 10^9$ —— ... and multiply the power by 10.

2. **Calculate $(2 \times 10^5) \div (4 \times 10^{10})$ without using a calculator. Give your answer in standard form.**

$(2 \times 10^5) \div (4 \times 10^{10})$

Divide front numbers and powers separately ——
$= \dfrac{2 \times 10^5}{4 \times 10^{10}} = \dfrac{2}{4} \times \dfrac{10^5}{10^{10}}$
$= 0.5 \times 10^{5-10}$ —— Subtract the powers (see p.28)

Not in standard form so convert it — multiply the number by 10...
$= 0.5 \times 10^{-5}$
$= 5 \times 10^{-6}$ —— ... and divide the power by 10.

Adding and Subtracting

1) Make sure the <u>powers of 10</u> are <u>the same</u>.
2) Add or subtract the <u>front numbers</u>.
3) Convert the answer to <u>standard form</u> if necessary.

To put standard form numbers into your calculator, use the [EXP] or the [×10ˣ] button.

E.g. enter 2.67×10^{15} by pressing

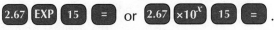

[2.67] [EXP] [15] [=] or [2.67] [×10ˣ] [15] [=] .

EXAMPLE

Calculate $(2.8 \times 10^4) + (6.6 \times 10^4)$ without using a calculator. Give your answer in standard form.

1) <u>Check</u> that both powers of 10 are equal.
$(2.8 \times 10^4) + (6.6 \times 10^4)$
2) Then add the <u>front numbers</u>.
$= (2.8 + 6.6) \times 10^4$
3) This is in standard form, so you <u>don't</u> need to convert it.
$= 9.4 \times 10^4$

Remember, n tells you how far the decimal point moves

Make sure you understand all the examples on these pages. Then answer these Exam Practice Questions:

Q1 Write a) 4.32×10^8 as an ordinary number. [1 mark]
   b) 0.000387 in standard form. [1 mark]

Q2 Which of these numbers is the smallest? 2.6×10^{-3} 1.9×10^{-3} 3.4×10^{-2} [1 mark]

Q3 Work out the following. Give your answers in standard form.
   a) $(9 \times 10^7) \div (3 \times 10^4)$ [3 marks]   b) $(6.7 \times 10^{10}) - (4.8 \times 10^{10})$ [2 marks]

Q3 Video Solution

Warm-up and Worked Exam Questions

I know that you'll be champing at the bit to get into the exam questions, but these warm-up questions are invaluable for getting the basic facts straight first.

Warm-up Questions

1) a) Use your calculator to find 0.8^4.
   b) Without a calculator, work out $3^4 - 2^5$.
   c) What is 1 000 000 as a power of 10?

2) Simplify: a) $4^1 \times 4^5$ b) $(5^3)^2$ c) $\dfrac{x^8}{x^2}$

3) Without using a calculator, find: a) $\dfrac{7^{12}}{7^3 \times 7^7}$ b) 3^{-3}

4) Find a) $\sqrt{225}$ b) $\sqrt[3]{6859}$ c) $\sqrt[5]{7776}$

5) The volume of a football is $\dfrac{4000\pi}{3}$ cm³.
   Given that the volume of a sphere = $\dfrac{4}{3}\pi r^3$, find the radius, r, of the ball in cm.

6) Work out $0.3^2(4.2 + 1.7)^3 + \sqrt{75.69}$ on your calculator. Write down your answer exactly.

7) Write a) 2 890 000 in standard form.
          b) 7.11×10^{-5} as an ordinary number.

8) Which of these numbers is the largest? 1.8×10^4 2.1×10^4 6.3×10^3

9) Work out the following. Give your answers in standard form.
   a) $(2 \times 10^{-4}) \times (6 \times 10^5)$ b) $(5.4 \times 10^{-7}) + (8.9 \times 10^{-7})$

Worked Exam Question

With the answers written in, it's very easy to just skim over this worked exam question.
But that's not really going to help you, so take the time to make sure you've really understood it.

1   The table on the right shows the masses of four different particles. (GRADE 5)

a) Which particle is the heaviest?

   Look at the power first.
   −6 moves the decimal point fewer
   places to the left than the others.

   ......C......
   [1 mark]

Particle	Mass (g)
Particle A	2.1×10^{-7}
Particle B	8.6×10^{-8}
Particle C	1.4×10^{-6}
Particle D	3.2×10^{-7}

b) What is the mass of particle C?
   Give your answer as an ordinary number.

   $1.4 \times 10^{-6} = 0.0000014$ g

   Move the decimal
   point 6 places.

   ......0.0000014...... g
   [1 mark]

c) How much more does particle D weigh than particle A?
   Give your answer in standard form.

   The power of 10 is the same
   for particles D and A, so it's
   just a simple subtraction.

   $(3.2 \times 10^{-7}) - (2.1 \times 10^{-7}) = (3.2 - 2.1) \times 10^{-7}$
   $= 1.1 \times 10^{-7}$ g

   ......1.1×10^{-7}...... g
   [2 marks]

Exam Questions

2   A square has an area of 6.25 cm². Find the length of one side of the square.   ③

.... cm
[1 mark]

3   $A = 4.834 \times 10^9$, $B = 2.4 \times 10^5$, $C = 5.21 \times 10^3$ ④

a)   Write A as an ordinary number.

....
[1 mark]

b)   Put A, B and C in order from smallest to largest.

.........., ..., ...
[1 mark]

4   Simplify the expression $\dfrac{3^4 \times 3^7}{3^6}$. Leave your answer in index form.   ⑤

....
[2 marks]

5   Work out the value of:

a) $6^5 \div 6^3$ ④

....
[1 mark]

b) $(2^4 \times 2^7) \div (2^3 \times 2^2)^2$ ⑤

....
[2 marks]

6   Light travels at approximately 2×10^5 miles per second.   ⑤
The distance from the Earth to the Sun is approximately 9×10^7 miles.

How long will it take light to travel this distance?

Use the formula:
time (s) = distance (miles) ÷ speed (miles/s)

.... seconds
[2 marks]

Revision Questions for Section One

That wraps up <u>Section One</u> — time to find out <u>what you know</u>.

- <u>Tick off each question</u> when you've <u>got it right</u>.
- When you're <u>completely happy</u> with a topic, tick that off too.

For even more practice, try the Sudden Fail Quiz for Section One — just scan this QR code!

Section One Quiz

Arithmetic (p1-6) ☑

Calculators are <u>only allowed</u> in questions 2, 7, 9, 10, 15 and 22. Sorry.

1) What are square numbers? Write down the first ten of them.

2) Using the numbers 2, 4 and 5, and +, −, × and ÷, what is the smallest possible positive number you can make? You can use each number/operation a maximum of once. You may also use brackets.

3) Tickets for a show cost £12 each. A senior's ticket is half price. A child's ticket is a third of the full price. How much does it cost for a family of 2 adults, 2 children and 1 senior to watch the show?

4) Find: a) £1.20 × 100   b) £150 ÷ 300

5) Work out: a) 51 × 27   b) 338 ÷ 13   c) 3.3 × 19   d) 4.2 ÷ 12

6) Find: a) −10 − 6   b) −35 ÷ −5   c) −4 + −5 + 22 − −7

Prime Numbers, Factors and Multiples (p9-12) ☑

7) Find all the prime numbers between 40 and 60 (there are 5 of them).

8) What are multiples? Find the first six multiples of: a) 10   b) 4

9) Express each of these as a product of prime factors: a) 210   b) 1050

10) Find: a) the HCF of 42 and 28   b) the LCM of 8 and 10

Fractions and Decimals (p15-19) ☑

11) Work out without a calculator: a) $\frac{25}{6} \div \frac{8}{3}$ b) $\frac{2}{3} \times 4\frac{2}{5}$ c) $\frac{5}{8} + \frac{9}{4}$ d) $\frac{2}{3} - \frac{1}{7}$

12) Calculate a) $\frac{4}{7}$ of 560   b) $\frac{2}{5}$ of £150

13) Amy, Brad and Cameron are all playing a video game. Amy has completed $\frac{5}{8}$ of the game, Brad has completed $\frac{7}{11}$ of the game and Cameron has completed $\frac{15}{22}$ of the game. Who has the largest fraction of the game left to complete?

14) Write: a) 0.04 as: (i) a fraction   (ii) a percentage   b) 65% as: (i) a fraction   (ii) a decimal

15) a) What is a recurring decimal?   b) Write $\frac{2}{9}$ as a recurring decimal.

Rounding and Estimating (p22-25) ☑

16) Round: a) 17.65 to 1 d.p. b) 6743 to 2 s.f. c) 3 643 510 to the nearest million.

17) Estimate the value of   a) $\frac{17.8 \times 32.3}{6.4}$ b) $\frac{96.2 \times 7.3}{0.463}$

18) Give the error interval for x if $x = 200$ when rounded to 1 s.f.

Powers and Roots (p28-29) ☑

19) If $f = 7^6 \times 7^4$ and $g = 7^5$, what is $f \div g$?

20) What is the value of 5^{-2}? Give your answer as a fraction.

21) Find without using a calculator: a) $\sqrt{121}$ b) $\sqrt[3]{64}$ c) $8^2 - 2^3$ d) 100 000 as a power of ten.

22) Use a calculator to find: a) 7.5^3 b) $\sqrt{23.04}$ c) $\sqrt[3]{512}$ d) $\sqrt[5]{161051}$

Standard Form (p30-31) ☑

23) What are the three rules for writing numbers in standard form?

24) Write: a) 3 560 000 000 in standard form b) 2.75×10^{-6} as an ordinary number.

25) Calculate: a) $(3.2 \times 10^6) \div (1.6 \times 10^3)$ b) $(5 \times 10^{11}) + (7 \times 10^{11})$
Give your answers in standard form.

Algebra — Simplifying

Algebra really terrifies so many people. But honestly, it's not that bad.
*Make sure you **understand and learn** these **basics** for dealing with algebraic expressions.*

Terms

Before you can do anything else with algebra, you must understand what a <u>term</u> is:

> A <u>**TERM**</u> is a collection of numbers, letters and brackets, all multiplied/divided together

Terms are separated by <u>+ and − signs</u>. Every term has a + or − attached to the <u>front of it</u>.

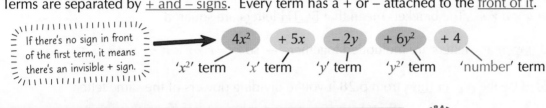

If there's no sign in front of the first term, it means there's an invisible + sign.

$4x^2$ $+5x$ $-2y$ $+6y^2$ $+4$

'x^2' term   'x' term   'y' term   'y^2' term   'number' term

Simplifying or 'Collecting Like Terms'

To <u>simplify</u> an algebraic expression made up of all the <u>same terms</u>, just <u>add</u> or <u>subtract</u> them.

EXAMPLES

1. Simplify $q + q + q + q + q$

'q' just means '1q'.

Just <u>add up</u> all the q's:
$q + q + q + q + q = 5q$

2. Simplify $4t + 5t - 2t$

Again, just <u>combine the terms</u> — don't forget there's a '−' before the 2t:

$4t + 5t - 2t = 7t$

If you have a mixture of <u>different terms</u>, it's a bit more tricky. To <u>simplify</u> an algebraic expression like this, you combine '<u>like terms</u>' (e.g. all the x terms, all the y terms, all the number terms etc.).

EXAMPLE   **Simplify** $2x - 4 + 5x + 6$

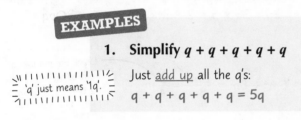

number terms

Invisible + sign

$(2x)\ (-4)\ (+5x)\ (+6) = (2x)\ (+5x)\ (-4)\ (+6) = 7x + 2$

x-terms $7x$ $+2$

1) Put <u>bubbles</u> round each term — be sure you capture the <u>+/− sign</u> in front of each.
2) Then you can move the bubbles into the <u>best order</u> so that <u>like terms</u> are together.
3) <u>Combine like terms</u>.

EXAMPLE   **Simplify** $4 + \sqrt{3} - 1 + 3\sqrt{3}$

Just treat $\sqrt{3}$-terms like x-terms — don't combine them with the number terms.

$\sqrt{3}$-terms

Invisible + sign

$(4)\ (+\sqrt{3})\ (-1)\ (+3\sqrt{3}) = (4)\ (-1)\ (+\sqrt{3})\ (+3\sqrt{3}) = 3 + 4\sqrt{3}$

number terms 3 $+4\sqrt{3}$

You need to know what terms are — and how to collect like ones

Terms are collections of numbers and letters separated by + and − signs. When you collect like terms together, you combine x terms, or y terms, or xy terms, or x^2 terms, or y^2 terms, or number terms, or...

Q1   Simplify: a) $2a + 7a + 4a$ [1 mark] (2)   b) $6b + 8b - 3b - b$ [1 mark] (2)

Q2   Simplify: a) $5x + y - 2x + 7y$ [1 mark] (2)   b) $6 + 5\sqrt{5} + 3 - 2\sqrt{5}$ [2 marks] (5)

Algebra — Multiplying and Dividing

*Multiplying algebra is a lot like multiplying numbers — here are a **few rules** to get you started.*

Multiplying and Dividing Letters

Watch out for these combinations of letters in algebra that regularly catch people out:

1) abc means $a \times b \times c$ and $3a$ means $3 \times a$. The ×'s are often left out to make it clearer.

2) gn^2 means $g \times n \times n$. Note that only the n is squared, not the g as well.

3) $(gn)^2$ means $g \times g \times n \times n$. The brackets mean that <u>BOTH</u> letters are squared.

4) <u>Powers</u> tell you <u>how many</u> letters are multiplied together — so $r^6 = r \times r \times r \times r \times r \times r$.

5) $\frac{a}{b}$ means $a \div b$. Use the <u>power rules</u> from p.28 if you're dividing powers of the same letter.

EXAMPLES

1. Simplify $k \times k \times k \times k$

You have 4 k's <u>multiplied together</u>:

$k \times k \times k \times k = k^4$

Careful — k times itself 4 times is k⁴, not 4k (4k means k + k + k + k or 4 × k).

2. Simplify $2p \times 3q \times 5$

Multiply the <u>numbers</u> together, then the <u>letters</u> together:

$2p \times 3q \times 5 = 2 \times 3 \times 5 \times p \times q = 30pq$

3. Simplify $6m^2 \div 8m$

Write as a <u>fraction</u> then simplify: $6m^2 \div 8m = \dfrac{6m^2}{8m} = \dfrac{3}{4}m$

$m^2 \div m = m$

Leave the number in front of the letter as a fraction not a decimal.

Multiplying Brackets

> The <u>key thing</u> to remember about multiplying brackets is that the thing <u>outside</u> the brackets multiplies <u>each separate term</u> inside the brackets.

EXAMPLE

Expand the following:

a) $3(2x + 5)$

$= (3 \times 2x) + (3 \times 5)$

$= 6x + 15$

b) $-4(3y - 2)$

$= (-4 \times 3y) + (-4 \times -2)$

$= -12y + 8$

c) $2e(e - 4)$

$= (2e \times e) + (2e \times -4)$

$= 2e^2 - 8e$

EXAMPLE

Expand $x(2x + 1) + y(y - 4) + 3x(y + 2)$

1) <u>Expand</u> each bracket separately.

$x(2x + 1) \quad + \quad y(y - 4) \quad + \quad 3x(y + 2)$

$= 2x^2 + x \quad + \quad y^2 - 4y \quad + \quad 3xy + 6x$

2) <u>Group together</u> like terms.

$= 2x^2 \quad + x + 6x \quad + 3xy \quad + y^2 \quad - 4y$

3) <u>Simplify</u> the expression.

$= 2x^2 + 7x + 3xy + y^2 - 4y$

Watch out when multiplying letters together

In the cases where multiplying and dividing letters gets a bit tricky, just remember the five points above. Remember the power rules — when multiplying, add powers, and when dividing, subtract powers.

Q1   Simplify the following expressions:

a) $5r \times -2s \times 6$ [1 mark]   b) $7(3m - 2)$ [1 mark]   c) $4p(p + 2q)$ [1 mark]

Q2   Show that $5(x + 8) + 2(x - 12) = 7x + 16$. [2 marks]

Q2 Video Solution

Inequalities

*Inequalities are a bit tricky, but once you've learned the tricks involved, most of the **algebra** for them is **identical** to ordinary **equations** (have a look back at pages 41-42 if you need a reminder).*

The **Inequality Symbols**

> > means 'Greater than'   ≥ means 'Greater than or equal to'
> < means 'Less than'   ≤ means 'Less than or equal to'

REMEMBER — the one at the BIG end is BIGGEST so $x > 4$ and $4 < x$ both mean: 'x is greater than 4'.

EXAMPLE   **x is an integer such that –4 < x ≤ 3. Write down all possible values of x.**

Work out what each bit of the inequality is telling you:

$-4 < x$ means 'x is greater than –4',

and $x \leq 3$ means 'x is less than or equal to 3'.

Remember, integers are just whole numbers (+ve and –ve, including 0).

Now just write down all the values that x can take: –3, –2, –1, 0, 1, 2, 3

–4 isn't included because of the < but 3 is included because of the ≤.

You Can Show Inequalities on **Number Lines**

Drawing inequalities on a number line is dead easy — all you have to remember is that you use an open circle (O) for > or < and a coloured-in circle (●) for ≥ or ≤.

EXAMPLE   **Show the inequality –4 < x ≤ 3 on a number line.**

Closed circle because 3 is included.

Open circle because –4 isn't included.

A number line from –5 to 5 with an open circle at –4 and a closed circle at 3.

Algebra with **Inequalities**

Solve inequalities like regular equations but WITH ONE BIG EXCEPTION:

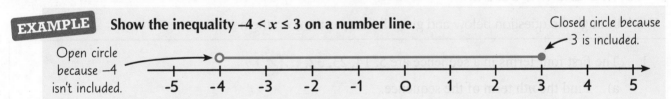

Whenever you MULTIPLY OR DIVIDE by a NEGATIVE NUMBER, you must FLIP THE INEQUALITY SIGN.

EXAMPLES

1. Solve 3x – 2 ≤ 13.

Just solve it like an equation — but leave the inequality sign in your answer:

(+2) $3x - 2 + 2 \leq 13 + 2$
          $3x \leq 15$
(÷3) $3x \div 3 \leq 15 \div 3$
          $x \leq 5$

2. Solve 2x + 7 > x + 11.

Again, solve it like an equation:

(–7) $2x + 7 - 7 > x + 11 - 7$
          $2x > x + 4$
(–x) $2x - x > x + 4 - x$
          $x > 4$

3. Solve 9 – 2x > 15.

Watch out for the sign change:

(–9) $9 - 2x - 9 > 15 - 9$
          $-2x > 6$
(÷–2) $-2x \div -2 < 6 \div -2$
          $x < -3$

The > has turned into a <, because we divided by a negative number.

Treat inequalities like equations

Learn the golden rules for solving equations and you'll be able to solve inequalities too. Just remember the extra rule about flipping the inequality sign if you multiply or divide by a negative number.

Q1 n is an integer such that $-1 \leq n < 5$. Write down all the possible values of n.   [2 marks]

Q2   Solve the following inequalities: a) $4x + 3 < 27$ [2 marks]   b) $4x \geq 18 - 2x$ [2 marks]

Warm-up and Worked Exam Questions

Sequences and inequalities — love them or hate them, you have to do them — it's just a case of learning the method and practising lots of questions. So let's start with some warm-up questions...

Warm-up Questions

1) Write down the next two terms in each of these sequences:
   a) 2, 6, 10, 14   b) 1, 3, 9, 27   c) 2, 3, 5, 8, 12

2) A sequence starts 1, 4, 9, 16, 25...
   a) What is the rule for this sequence?
   b) Find the next two terms of the sequence.
   c) Is 100 a term in this sequence? Explain your answer.

3) For each of the sequences, say whether it is an arithmetic or geometric sequence and explain why.   a) 5, 9, 13, 17...   b) 4, 12, 36, 108...

4) A sequence starts 27, 23, 19, 15...
   a) Find an expression for the nth term of the sequence.
   b) What is the 20th term in the sequence?

5) n is an integer such that $-3 < n \leq 4$. Write down all the possible values of n.

6) Solve these inequalities: a) $6x - 2 < 28$ b) $5x \geq 49 - 2x$

Worked Exam Question

Work through the question below and give all the questions on the next page a good go.

1   The first four terms in a sequence are 5, 14, 23, 32, …

   a)   Find the nth term of the sequence.

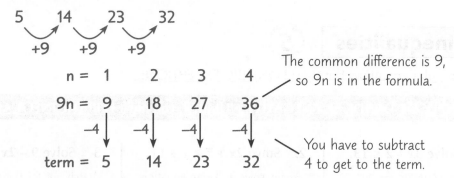

The common difference is 9, so 9n is in the formula.

You have to subtract 4 to get to the term.

So the expression for the nth term is 9n − 4

                                                                                          9n − 4
                                                                                    ....
                                                                                         [2 marks]

   b)   What is the 20th term of the sequence?

         20th term = (9 × 20) − 4 = 176

                                                                                          176
                                                                                    ....
                                                                                         [1 mark]

   c)   Is 87 a term in this sequence? Explain your answer.

         If 87 is a term in the sequence, then 9n − 4 = 87, 9n = 91 so n = 10.111…

         n is not a whole number, so 87 is not a term in the sequence.

                                                                                         [2 marks]

Exam Questions

2 n is an integer. List all the possible values of n that satisfy the inequality $-3 \leq n < 2$. **(3)**

....

[2 marks]

3 The patterns in the sequence below represent the first three triangle numbers. **(3)**

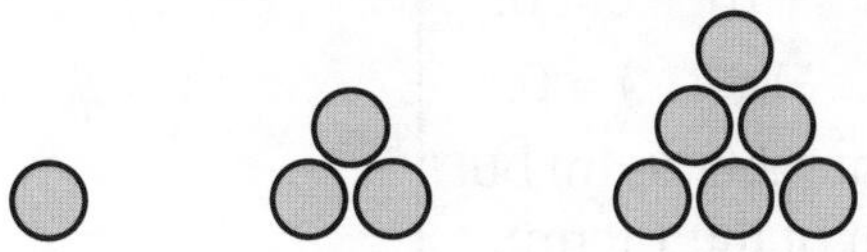

a) Draw the next pattern in the sequence.

[1 mark]

b) How many circles are in the tenth pattern in the sequence? Give a reason for your answer.

....

....

[2 marks]

4 To find the next term in the sequence below, you add together the two previous terms. **(4)**
Fill in the gaps to complete the sequence.

| 3 | .... | 7 | .... | .... | 29 |

[2 marks]

5 A quadratic sequence starts 2, 6, 12, 20, … **(4)**
Find the next term in the sequence.

Find the pattern in the differences between each pair of terms and use this to find the next term.

....

[2 marks]

6 p and q are integers. $p \leq 45$ and $q > 25$. **(4)**
What is the largest possible value of $p - q$?

....

[2 marks]

Section Two — Algebra

Quadratic Equations

A **quadratic** equation is one where the highest power is x^2.
The standard format for a quadratic equation is $x^2 + bx + c = 0$.

You can **Factorise Quadratic Equations**

1) You can <u>solve</u> quadratic equations by <u>factorising</u>.
2) '<u>Factorising a quadratic</u>' means '<u>putting it into 2 brackets</u>'.

Factorising Quadratics

1) <u>ALWAYS</u> rearrange into the <u>STANDARD FORMAT</u>: $x^2 + bx + c = 0$.
2) Write down the <u>TWO BRACKETS</u> with the x's in: $(x\quad)(x\quad) = 0$:
3) Then <u>find 2 numbers</u> that <u>MULTIPLY to give 'c'</u> (the number term) but also <u>ADD/SUBTRACT to give 'b'</u> (the number in front of the x term).
4) Fill in the +/− signs and make sure they work out properly.
5) As an <u>ESSENTIAL CHECK</u>, <u>EXPAND</u> the brackets to make sure they give the original equation.

Ignore any minus signs at this stage.

3) As well as factorising a quadratic, you might be asked to <u>solve</u> the equation. This just means finding the values of x that make each bracket <u>0</u> (see example below).

EXAMPLE   Solve $x^2 - x = 12$.

1) $x^2 - x - 12 = 0$
    $(b = -1, c = -12)$

1) <u>Rearrange</u> into the standard format.

2) $(x\quad)(x\quad) = 0$

2) Write down <u>two brackets</u> with x's in.

3)
| 1×12 Add/subtract to give: 13 or 11 |
| 2×6 Add/subtract to give: 8 or 4 |
| 3×4 Add/subtract to give: 7 or ①|

$(x\quad 3)(x\quad 4) = 0$ This is what we want.

3) Find the right <u>pair of numbers</u> that <u>multiply to give c</u> (= 12), and <u>add or subtract to give b</u> (= 1) (remember, we're ignoring the +/− signs for now).

4) $(x + 3)(x - 4) = 0$

4) <u>Now fill in the +/− signs</u> so that 3 and 4 add/subtract to give −1 (= b).

5) Check:
   $(x + 3)(x - 4) = x^2 - 4x + 3x - 12$
   $\qquad\qquad\qquad = x^2 - x - 12$ ✓

5) <u>ESSENTIAL check</u> — <u>EXPAND the brackets</u> to make sure they give the original expression.

But we're not finished yet — we've only factorised it, we still need to...

6) $(x + 3) = 0 \Rightarrow x = -3$
    $(x - 4) = 0 \Rightarrow x = 4$

6) <u>SOLVE THE EQUATION</u> by setting each bracket <u>equal to 0</u>.

To help you work out which <u>signs</u> you need, <u>look at c</u>.

• If c is <u>positive</u>, the signs will be <u>the same</u> — both positive or both negative.
• If c is <u>negative</u> the signs will be <u>different</u> — one positive and one negative.

Factorising quadratics is not easy — but it is important

Make sure you learn the method, then it's a case of practise, practise, practise. People often get the signs the wrong way round, so make sure you always check your answer by expanding the brackets.

Q1   Factorise $x^2 + 2x - 15$ [2 marks]

Q2   Solve $x^2 - 9x + 20 = 0$ [3 marks]

Q2 Video Solution

Simultaneous Equations

Simultaneous equations might sound a bit scary, but they're just a **pair** of equations that you have to solve **at the same time**. You have to find values of x and y that work in **both** equations.

Six Steps for **Simultaneous Equations**

EXAMPLE   Solve the simultaneous equations $2x + 4y = 6$
$4x + 3y = -3$

Your equations should be in the form $ax + by = c$, where a, b and c are numbers, so rearrange them if you need to.

1 <u>Label</u> your equations ① and ②.

$2x + 4y = 6$ — ①
$4x + 3y = -3$ — ②

2 <u>Match up the numbers in front</u> of either the x's or y's in both equations.
You may need to multiply one or both equations by a suitable number.
Relabel the equations ③ and ④ if you need to change them.

① × 2: $4x + 8y = 12$ — ③
$4x + 3y = -3$ — ②

You don't need to change equation 2 for this example.

3 <u>Add or subtract the two equations</u> to eliminate the terms with the same number in front.

③ − ②: $4x + 8y = 12$
$- \quad 4x + 3y = -3$
$\overline{0x + 5y = 15}$

If the numbers have the <u>same sign</u> (both +ve or both −ve) then <u>subtract</u>. If the numbers have <u>opposite signs</u> (one +ve and one −ve) then <u>add</u>.

4 <u>Solve</u> the resulting equation.

$5y = 15 \implies \underline{y = 3}$

5 <u>Substitute</u> the value you've found back into equation ① and <u>solve it</u>.

Sub $y = 3$ into ①: $2x + (4 \times 3) = 6 \implies 2x + 12 = 6 \implies 2x = -6 \implies \underline{x = -3}$

6 <u>Substitute</u> both these values into equation ② to make sure it works.
If it doesn't then you've done something wrong and you'll have to do it all again.

Sub x and y into ②: $(4 \times -3) + (3 \times 3) = -12 + 9 = -3$, which is right, so it's worked.
So the solutions are: $x = -3, \quad y = 3$

It doesn't matter if you eliminate the x's or y's... do whatever's easiest

It might just be me, but I think simultaneous equations are quite fun... well, maybe not fun... but quite satisfying. Anyway, it doesn't matter whether you like them or not — you have to learn how to do them.

Q1   Solve the simultaneous equations $x + y = 3$
$4x + 3y = 10$ [3 marks]

Q2   Solve the simultaneous equations $2x + 4y = 2$
$5x - 3y = 18$ [3 marks]

Proof

*I'm not going to lie — **proof questions** can look a bit terrifying.*
But there are a couple of tricks you can use that make them a bit less scary.

Prove Statements are True or False

1) The most straightforward proofs are ones where you're given a <u>statement</u> and asked if it's <u>true</u> or <u>false</u>.

2) To show that it's <u>false</u>, all you have to do is find <u>one example</u> that doesn't <u>work</u>.

3) Showing that something is <u>true</u> is a bit trickier — you might have to do a bit of <u>rearranging</u> to show that two things are <u>equal</u>, or show that one thing is a <u>multiple</u> of a certain number.

 Find an example to show that the statement below is not correct.
"The difference between two prime numbers is always even."

2 and 5 are both prime, so try them:
5 − 2 = 3, which is odd —
so the statement is not correct.

> It was easy to find an example for this one — but sometimes you might have to try a few different numbers to find a pair that doesn't work.

 Prove that $(n + 3)^2 - (n - 2)^2 \equiv 5(2n + 1)$.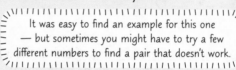

Take one side of the equation and play about with it until you get the other side:

LHS: $(n + 3)^2 - (n - 2)^2 \equiv n^2 + 6n + 9 - (n^2 - 4n + 4)$
$\equiv n^2 + 6n + 9 - n^2 + 4n - 4$
$\equiv 10n + 5$
$\equiv 5(2n + 1) = $ RHS ✓

> See p.38 for a reminder on factorising.

> $\equiv$ is the <u>identity symbol</u>, and means that two things are <u>identically equal</u> to each other. So $a + b \equiv b + a$ is true for <u>all values</u> of a and b (unlike an equation, which is only true for certain values).

Show that One Thing is a Multiple of Another

1) To show that something is a <u>multiple</u> of a particular number (let's say <u>5</u>), you need to <u>rearrange</u> the thing you're given to get it into the form <u>5 × a whole number</u>, which means it's a multiple of 5.

2) If it <u>can't</u> be written as 5 × a whole number, then it's <u>not</u> a multiple of 5.

EXAMPLE $a = 3(b + 9) + 5(b - 2) + 3$.
Show that a is a multiple of 4 for any whole number value of b.

$a = 3(b + 9) + 5(b - 2) + 3$
$= 3b + 27 + 5b - 10 + 3$ —— Expand the brackets...
$= 8b + 20$ —————— ... simplify...
$= 4(2b + 5)$ ————— ... and factorise.

> $2b + 5$ is a whole number because b is a whole number.

a can be written as 4 × something (where the something is 2b + 5) so it is a multiple of 4.

3) It's always a good idea to keep in mind what you're <u>aiming for</u> — here, you're trying to write the expression for a as '<u>4 × a whole number</u>', so you'll need to take out a <u>factor of 4</u> at some point.

Proof questions aren't as bad as they look

If you're asked to prove that two things are equal, rearranging should do the trick. If you're asked to prove something is wrong, just find an example of it not working. <u>Always</u> keep in mind what you're aiming for.

Q1 For each of the following statements, find an example to prove that they are false.
  a) All square numbers end in 1, 4, 6 or 9. [1 mark]
  b) The product of two prime numbers is always odd. [1 mark]

Warm-up and Worked Exam Questions

A mixed bag of tricky topics to end the algebra section — the only way you're going to get your head around them is to do some practice questions. Start by warming up with these questions.

Warm-up Questions

1) Factorise the following expressions:
   a) $x^2 - 6x + 8$ b) $x^2 - 2x - 3$ c) $x^2 + 7x + 12$

2) Solve the following equations:
   a) $x^2 + 7x + 10 = 0$ b) $x^2 + 5x - 14 = 0$ c) $x^2 - 5x + 3 = -3$

3) Solve the simultaneous equations: $x + y = 5$ and $2y - 3x = 5$.

4) Solve the simultaneous equations: $3x + 5y = 12$ and $4x - 3y = -13$.

5) Give an example to prove that each of the following statements is false.
   a) The product of two square numbers is never square.
   b) The sum of two prime numbers is always prime.

6) Prove that $(x + 2)^2 + (x - 2)^2 = 2(x^2 + 4)$ for all values of x.

Worked Exam Questions

That's enough warming up, it's time for the real thing. Take a look at these worked examples first and then turn over and have a go at some practice questions yourself.

1 The equation $x^2 + x - 20 = 0$ is an example of a quadratic equation.

     a) Fully factorise the expression $x^2 + x - 20$.

Number pairs that multiply to give 20 are: 5 and −4 add to give 1
1×20, 2×10, 4×5
$$x^2 + x - 20 = (x + 5)(x - 4)$$

$(x + 5)(x - 4)$
....
[2 marks]

     b) Use your answer to part a) to solve the equation $x^2 + x - 20 = 0$.

$(x + 5)(x - 4) = 0$
So, $x + 5 = 0$ or $x - 4 = 0$
    $x = -5$ or $x = 4$

$x = $ ....−5.... or $x = $ ....4....
[1 mark]

2 Solve this pair of simultaneous equations.

$4x + 3y = 16$ ①   ① − ②: $4x + 3y = 16$ Eliminate the x term by
$4x + 2y = 12$ ② $- 4x + 2y = 12$ subtracting the equations
                         $y = 4$

Substitute $y = 4$
into equation ①. $4x + (3 \times 4) = 16$
               $4x = 4$ so $x = 1$

Check your answer by putting
$x = 1$ and $y = 4$ into ② — $(4 \times 1) + (2 \times 4) = 12$ ✓

$x = $ ....1.... $y = $ ....4....
[2 marks]

Exam Questions

3   For each statement below, write down an example to show that the statement is incorrect. **(4)**

a)   There are no factors of 48 between 15 and 20.

....
[1 mark]

b)   The sum of two square numbers is always odd.

....
[1 mark]

c)   All numbers that end in an 8 are multiples of either 4, 6 or 8.

....
[1 mark]

4   Solve this pair of simultaneous equations. **(5)**

$x + 3y = 11$
$3x + y = 9$

$x = $ .... $y = $ ....
[3 marks]

5   Solve the equation $x^2 + 4x - 12 = 0$. **(5)**

$x = $ ....   or $x = $ ....
[3 marks]

6   q is a whole number. Show that $2(18 + 3q) + 3(3 + q)$ is a multiple of 9. **(5)**

[3 marks]

Revision Questions for Section Two

Now, let's see how much of that <u>tricky algebra</u> you've learned.

- Try these questions and <u>tick off each one</u> when you <u>get it right</u>.
- When you're done <u>completely happy</u> with it, tick off the topic.

For even more practice, try the Sudden Fail Quiz for Section Two — just scan this QR code!

Section Two Quiz

Algebra (p35-38) ☑

1) Simplify: a) $e + e + e$ b) $6f + 7f - f$

2) Simplify: a) $2x + 3y + 5x - 4y$ b) $11a + 2 - 8a + 7$

3) Simplify: a) $m \times m \times m$ b) $p \times q \times 7$ c) $2x \times 9y$

4) Expand: a) $6(x + 3)$ b) $-3(3x - 4)$ c) $x(5 - x)$

5) Expand and simplify $4(3 + 5x) - 2(7x + 6)$

6) Expand and simplify: a) $(x + 2)(2x - 5)$ b) $(5y + 2)^2$

7) What is factorising?

8) Factorise: a) $8x + 24$ b) $18x^2 + 27x$ c) $36x^2 - 81y^2$

Solving Equations (p41-42) ☑

9) Solve: a) $x + 9 = 16$ b) $x - 4 = 12$ c) $6x = 18$

10) Solve: a) $4x + 3 = 19$ b) $3x + 6 = x + 10$ c) $3(x + 2) = 5x$

Expressions, Formulas and Functions (p43-46) ☑

11) $Q = 5r + 6s$. Work out the value of Q when $r = -2$ and $s = 3$.

12) A function takes a number, doubles it and subtracts 8.
    What is the result when 11 is put in the machine?

13) Tony and Robbie have the same number of marbles. Nadia has 26 marbles.
    Between them, they have 100 marbles. How many marbles does Tony have?

14) A rectangle measures $2x$ cm by $7x$ cm. An equilateral triangle has the same perimeter
    as the rectangle. Find the length of one side of the triangle in terms of x.

15) Rearrange the formula $W = 4v + 5$ to make v the subject.

Sequences and Inequalities (p49-51) ☑

16) For each of the following sequences, find the next term and write down the rule you used.
    a) 3, 10, 17, 24, ...   b) 1, 4, 16, 64, ...   c) 2, 5, 7, 12, ...

17) Find an expression for the nth term of the sequence that starts 4, 10, 16, 22, ...

18) Is 34 a term in the sequence given by the expression $7n - 1$?

19) Write the following inequalities out in words: a) $x > -7$ b) $x \leq 6$

20) $0 < k \leq 7$. Find all the possible integer values of k.

21) Solve the following inequalities: a) $x + 4 < 14$ b) $3x + 5 \leq 26$.

Quadratic Equations (p54) ☑

22) Factorise the following quadratic expressions: a) $x^2 + 10x + 16$ b) $x^2 - 6x - 7$

23) Solve the quadratic equation $x^2 + 3x - 18 = 0$.

Simultaneous Equations and Proof (p55-56) ☑

24) Solve the following pair of simultaneous equations: $4x + 5y = 33$ and $-x + 3y = 13$

25) Find an example to show that the statement
    "if a number ends in a 9, it must divide by 3" is not true.

26) Show that x is never a multiple of 5 if $x = 3(y + 2) + 2(y + 6)$ for any whole number y.

Coordinates and Midpoints

*What could be more fun than points in one quadrant? Points in **four quadrants**, that's what...*

The Four **Quadrants**

A graph has <u>four different quadrants</u> (regions).

The top-right region is the easiest because <u>ALL THE COORDINATES IN IT ARE POSITIVE</u>.

You have to be careful in the <u>other regions</u> though, because the *x*- and *y*- coordinates could be <u>negative</u>, and that makes life much more difficult.

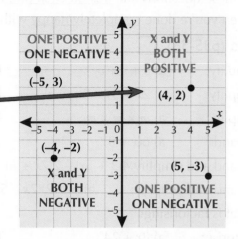

THREE IMPORTANT POINTS ABOUT COORDINATES:

1) The coordinates are always in <u>ALPHABETICAL ORDER</u>, *x* then *y*. (x, y)

2) *x* is always the flat axis going <u>ACROSS</u> the page. In other words '*x* is a...cross'.

3) Remember it's always <u>IN THE HOUSE</u> (→) and then <u>UP THE STAIRS</u> (↑) so it's <u>ALONG first</u> and <u>then UP</u>, i.e. *x*-coordinate first, and then *y*-coordinate

The **Midpoint** of a Line

The '<u>MIDPOINT OF A LINE SEGMENT</u>' is the <u>POINT THAT'S BANG IN THE MIDDLE</u> of it.

Finding the coordinates of a midpoint is pretty easy.
<u>LEARN THESE THREE STEPS</u>...

> 1) Find the <u>average</u> of the <u>*x*-coordinates</u>.
> 2) Find the <u>average</u> of the <u>*y*-coordinates</u>.
> 3) Plonk them in <u>brackets</u>.

EXAMPLE

P and Q have coordinates (1, 2) and (6, 6). Find the midpoint of the line PQ.

Average of *x*-coordinates = $\frac{1+6}{2}$ = 3.5

Average of *y*-coordinates = $\frac{2+6}{2}$ = 4

Coordinates of midpoint = (3.5, 4)

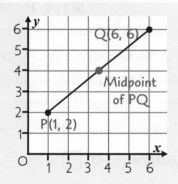

Coordinates should always be written as (x, y)

Learn the three points for getting *x* and *y* the right way round and the three easy steps for finding the midpoint of a line segment. Then, put your skills to the test with these Exam Practice Questions.

Q1   a) Plot point A(–3, 2) and point B(3, 5) on a grid.   [2 marks]

      b) Find the coordinates of the midpoint of AB.   [2 marks]

Q1 Video Solution

Straight-Line Graphs

*Here are the basic straight-line graphs — you need to be able to **draw them** and give their **equations**.*

Vertical and Horizontal lines: 'x = a' and 'y = a'

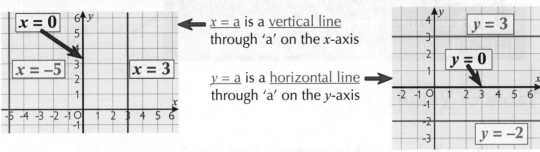

$x = a$ is a <u>vertical line</u> through 'a' on the x-axis

$y = a$ is a <u>horizontal line</u> through 'a' on the y-axis

The Main Diagonals: 'y = x' and 'y = −x'

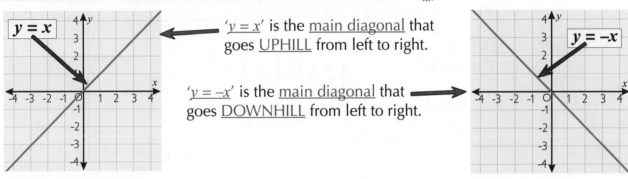

'$y = x$' is the <u>main diagonal</u> that goes <u>UPHILL</u> from left to right.

'$y = -x$' is the <u>main diagonal</u> that goes <u>DOWNHILL</u> from left to right.

Other Lines Through the Origin: 'y = ax' and 'y = −ax'

$y = ax$ and $y = -ax$ are the equations for **A SLOPING LINE THROUGH THE ORIGIN.**

The value of '<u>a</u>' (known as the <u>gradient</u>) tells you the steepness of the line. The bigger 'a' is, the steeper the slope. A <u>MINUS SIGN</u> tells you it slopes <u>DOWNHILL</u>.

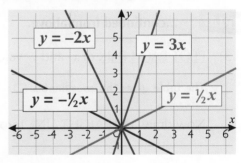

Learn to Spot Straight Lines from their Equations

All straight-line equations just contain '<u>something x</u>, <u>something y</u> and <u>a number</u>'.

<u>Straight lines:</u>		<u>NOT straight lines:</u>	
$x - y = 0$	$y = 2 + 3x$	$y = x^3 + 3$	$\frac{1}{y} + \frac{1}{x} = 2$
$2y - 4x = 7$	$4x - 3 = 5y$	$x^2 = 4 - y$	$xy + 3 = 0$

There's more on x^2 graphs on page 68.

Get it straight — which lines are straight (and which aren't)

It's definitely worth learning all the graphs above. Once you've done that, test yourself with this question.

Q1 On a grid with x-axis from −5 to 5 and y-axis from −5 to 5, draw these lines:

a) $y = -1$ b) $y = -x$ c) $x = 2$ [3 marks]

Drawing Straight-Line Graphs

*You might be asked to **DRAW THE GRAPH** of an equation in the exam.*
*This **EASY METHOD** will net you the marks every time:*

> 1) Choose 3 values of x and <u>draw up a table</u>.
> 2) <u>Work out the corresponding y-values.</u>
> 3) <u>Plot the coordinates</u>, and <u>draw the line</u>.

You might get lucky and be <u>given</u> a table in an exam question. Don't worry if it contains <u>5 or 6 values</u>.

Doing the 'Table of Values'

EXAMPLE   **Draw the graph of $y = 2x - 3$ for values of x from -2 to 4.**

1) <u>Choose 3 easy x-values for your table:</u>
   Use x-values from the grid you're given.
   Avoid negative ones if you can.

x	O	2	4
y			

2) <u>Find the y-values</u> by putting each x-value into the equation:

x	O	2	4
y	-3	1	5

   When $x = $ O,
   $y = 2x - 3$
   $\quad = (2 \times \text{O}) - 3 = -3$

   When $x = 4$,
   $y = 2x - 3$
   $\quad = (2 \times 4) - 3 = 5$

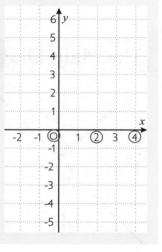

Plotting the Points and Drawing the Graph

EXAMPLE   **...continued from above.**

3) <u>PLOT EACH PAIR</u> of x- and y- values from your table.

   The table gives the coordinates (O, -3), (2, 1) and (4, 5).

   Now draw a <u>STRAIGHT LINE</u> through your points.

   If one point looks a bit wacky, check 2 things:
   – the <u>y-values</u> you worked out in the table
   – that you've <u>plotted</u> the points properly.

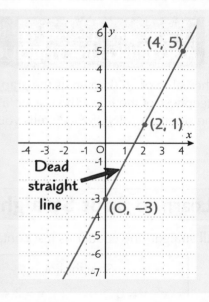

Dead straight line

Spot and plot a straight line — then check it looks right

In the exam you might get an equation like $3x + y = 5$ to plot, making finding the y-values a bit trickier.
Just substitute in the x-value and find the y-value that makes the equation true.
E.g. when $x = 1$, $3x + y = 5 \rightarrow (3 \times 1) + y = 5 \rightarrow 3 + y = 5 \rightarrow y = 2$.

Q1   Draw the graph of $y = x + 4$ for values of x from -6 to 2.   [3 marks]

Q2   Draw the graph of $y + 3x = 2$ for values of x from -2 to 2.   [3 marks]

Q2 Video Solution

Straight-Line Graphs — Gradients

Time to hit the slopes. Well, find them anyway...

Finding the **Gradient**

The gradient of a line is a measure of its slope. The bigger the number, the steeper the line.

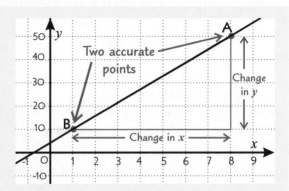

EXAMPLE **Find the gradient of the straight line shown.**

1 Find two accurate points and complete the triangle.

Choose easy points with positive coordinates.

Two points that can be read accurately are:

Point A: (8, 50)   Point B: (1, 10)

2 Find the change in y and the change in x.

Change in y = 50 − 10 = <u>40</u>

Change in x = 8 − 1 = <u>7</u>

Make sure you subtract the x-coordinates the SAME WAY ROUND as you do the y-coordinates. E.g. y-coord. of pt A − y-coord. of pt B and x-coord. of pt A − x-coord. of pt B

3 LEARN this formula, and use it:

$$\text{GRADIENT} = \frac{\text{CHANGE IN Y}}{\text{CHANGE IN X}}$$

Gradient = $\frac{40}{7}$ = <u>5.71</u> (to 2 d.p.)

Make sure you get the formula the right way up. Remember it's <u>VER</u>y <u>HO</u>t — <u>VER</u>tical over <u>HO</u>rizontal.

4 Check the sign's right.

If it slopes uphill left → right (⟋) then it's positive.
If it slopes downhill left → right (⟍) then it's negative.

As the graph goes uphill, the gradient is positive. So the gradient is <u>5.71</u> (not -5.71).

1) In real life, gradients are often given as a ratio or a percentage.
2) You'll often see them on road signs to describe the steepness of hills and slopes.
3) E.g. if the gradient = $\frac{1}{4}$, it has a ratio of <u>1 : 4</u>, a percentage of <u>25%</u> and you say it's "<u>1 in 4</u>" (for every 4 units you move horizontally, you move 1 unit vertically).

Gradient = change in y over change in x

Learn the four steps for finding a gradient then have a shot at this Exam Practice Question. Take care — you might not be able to pick two points with nice, positive coordinates.

Q1   Find the gradient of the line shown.   [2 marks]

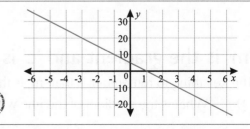

Straight-Line Graphs — y = mx + c

This sounds a bit scary, but give it a go and you might like it.

y = mx + c is the Equation of a **Straight Line**

$y = mx + c$ is the general equation for a straight-line graph, and you need to remember:

> 'm' is equal to the <u>GRADIENT</u> of the graph, 'c' is the value <u>WHERE IT CROSSES THE Y-AXIS</u> and is called the <u>Y-INTERCEPT</u>.

'<u>m</u>' and '<u>c</u>' are always just <u>numbers</u> — so $y = 3x - 1$ and $y = -x + 2$ are in $y = mx + c$ form.

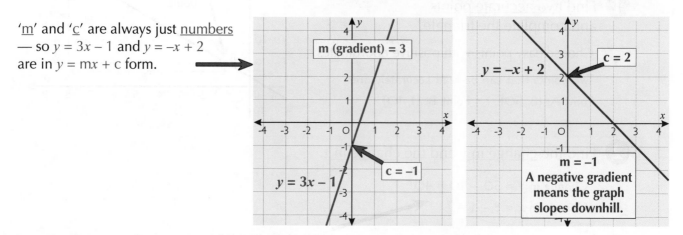

You might have to <u>rearrange</u> a straight-line equation to get it into this form:

Straight line:		Rearranged into '$y = mx + c$'	
$y = 2 + 3x$	$\rightarrow$	$y = 3x + 2$	(m = 3, c = 2)
$x - y = 4$	$\rightarrow$	$y = x - 4$	(m = 1, c = −4)
$4 - 3x = y$	$\rightarrow$	$y = -3x + 4$	(m = −3, c = 4)

<u>WATCH OUT</u>: people mix up 'm' and 'c' when they get something like $y = 5 + 2x$. Remember, 'm' is the number <u>in front of the 'x'</u> and 'c' is the number <u>on its own</u>.

Finding the **Equation** of a Straight-Line **Graph**

EXAMPLE   **Find the equation of the line on the graph in the form y = mx + c.**

1 Find '<u>m</u>' (gradient) $m = \dfrac{\text{change in } y}{\text{change in } x} = \dfrac{15}{30} = \dfrac{1}{2}$

It's an uphill graph, so the gradient is positive.

2 Read off '<u>c</u>' (y-intercept) $c = 15$

3 Use these to write the equation in the form $y = mx + c$. $y = \dfrac{1}{2}x + 15$

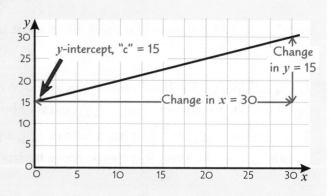

'm' is the gradient and 'c' is the y-intercept

The key thing to remember is that 'm' is the number in front of the x, and 'c' is the number on its own.

Q1   What is the gradient of the line with equation $y = 4 - 2x$?   [1 mark]

Section Three — Graphs

Using y = mx + c

This page covers some of the awkward questions you might get asked about straight lines.

Parallel Lines Have the Same Gradient

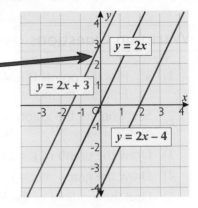

Parallel lines all have the <u>same gradient</u>, which means their
$y = mx + c$ equations all have the same value of <u>m</u>.
So the lines: $y = 2x + 3$, $y = 2x$ and $y = 2x - 4$ are all parallel.

EXAMPLE   **Line J has a gradient of –3. Find the equation of Line K, which is parallel to Line J and passes through point (2, 3).**

Lines J and K are <u>parallel</u> so their <u>gradients</u>
are the same $\Rightarrow$ m = –3

$y = -3x + c$

When $x = 2$, $y = 3$:
$3 = (-3 \times 2) + c \Rightarrow 3 = -6 + c$
$c = 9$

$y = -3x + 9$

1) First find the <u>'m' value</u> for Line K.

2) Substitute the value for 'm' into <u>$y = mx + c$</u> to give you the 'equation so far'.

3) Substitute the <u>x and y values</u> for the given point on Line K and solve for '<u>c</u>'.

4) Write out the <u>full equation</u>.

Finding the Equation of a Line Through Two Points

If you're given <u>two points</u> on a line you can find the <u>gradient</u>, then you can <u>use</u> the gradient and one of the points to find the <u>equation</u> of the line. It's a bit <u>tricky</u>, but try to follow the <u>method</u> used in this example.

EXAMPLE   **Find the equation of the straight line that passes through (–2, 9) and (3, –1).**
Give your answer in the form $y = mx + c$.

1) Use the <u>two</u> points to find '<u>m</u>' (gradient).

$m = \dfrac{\text{change in } y}{\text{change in } x} = \dfrac{-1-9}{3-(-2)} = \dfrac{-10}{5} = -2$

So $y = -2x + c$

2) <u>Substitute</u> one of the points into the equation you've just found.

Substitute (–2, 9) into eqn: $9 = (-2 \times -2) + c$
$9 = 4 + c$

3) <u>Rearrange</u> the equation to find '<u>c</u>'.

$c = 9 - 4$
$c = 5$

4) Write out the <u>full equation</u>.

$y = -2x + 5$

Sometimes you'll be asked to give your equation in other forms, such as $ax + by + c = 0$.
Just <u>rearrange</u> your $y = mx + c$ equation to get it in this form.

Parallel lines have the same gradient

To check that lines are parallel, rearrange each equation into the form $y = mx + c$ and then compare their values of m — if they're the same then the lines are parallel, if they're different then they aren't parallel.

Q1   a) Line Q goes through (0, 5) and (4, 7).
Find the equation of Line Q in the form $y = mx + c$.   [3 marks]
b) Line R is parallel to line Q. It intersects the y-axis at (0, 10).
Write down the equation of line R in the form $y = mx + c$.   [1 mark]

Q1 Video Solution

Ratios

*Another page on **ratios** coming up — it's more **interesting** than the first but not as exciting as the next one...*

Scaling Up Ratios

If you know the ratio between parts and the actual size of one part,
you can scale the ratio up to find the other parts.

> **EXAMPLE**
> **Mortar is made from mixing sand and cement in the ratio 7:2. How many buckets of mortar will be made if 21 buckets of sand are used in the mixture?**
>
> You need to multiply by 3 to go from 7 to 21 on the left-hand side (LHS) — so do that to both sides:
>
> So 21 buckets of sand and 6 buckets of cement are used.
>
> sand:cement
> $\times 3 \big(\begin{smallmatrix} 7:2 \\ 21:6 \end{smallmatrix} \big) \times 3$
>
> Amount of mortar made = 21 + 6 = 27 buckets

The two parts of a ratio are always in direct proportion (see p.83-84). So in the example above, sand and cement are in direct proportion, e.g. if the amount of sand doubles, the amount of cement doubles.

Writing Ratios as Fractions

1) To write one part as a fraction of another part — put one number over the other.

   E.g. if apples and oranges are in the ratio 2:9 then we say there are $\frac{2}{9}$ as many apples as oranges or $\frac{9}{2}$ times as many oranges as apples.

2) To write one part as a fraction of the total — add up the parts to find the total, then put the part you want over the total.

   E.g. a pie dough is made by mixing flour, butter and water in the ratio 3:2:1.
   The total number of parts is $3+2+1=\underline{6}$.
   So $\frac{3}{6} = \frac{1}{2}$ of the dough is flour, $\frac{2}{6} = \frac{1}{3}$ is butter and $\frac{1}{6}$ is water.

Part : Whole Ratios

You might come across a ratio where the LHS is included in the RHS — these are called part:whole ratios.

> **EXAMPLE**
> **Mrs Miggins owns tabby cats and ginger cats.
> The ratio of tabby cats to the total number of cats is 3:5.**
>
> **a) What fraction of Mrs Miggins' cats are tabby cats?**
> The ratio tells you that for every 5 cats, 3 are tabby cats.
> $\frac{\text{part}}{\text{whole}} = \frac{3}{5}$
>
> **b) What is the ratio of tabby cats to ginger cats?**
> 3 in every 5 cats are tabby, so 2 in every 5 are ginger.
> $5 - 3 = 2$
> For every 3 tabby cats there are 2 ginger cats.
> tabby:ginger = 3:2
>
> **c) Mrs Miggins has 12 tabby cats. How many ginger cats does she have?**
> Scale up the ratio from part b) to find the number of ginger cats.
> tabby:ginger
> $\times 4 \big(\begin{smallmatrix} 3:2 \\ 12:8 \end{smallmatrix} \big) \times 4$
> There are 8 ginger cats

Ratios

*If you were worried I was running out of **great stuff** to say about ratios then worry no more...*

Proportional **Division**

In a proportional division question a TOTAL AMOUNT is split into parts in a certain ratio.
The key word here is PARTS — concentrate on 'parts' and it all becomes quite painless:

EXAMPLE   **Jess, Mo and Suria share £9100 in the ratio 2:4:7. How much does Mo get?**

1) ADD UP THE PARTS:
   The ratio 2:4:7 means there will be a total of 13 parts: 2 + 4 + 7 = 13 parts

2) DIVIDE TO FIND ONE "PART":
   Just divide the total amount by the number of parts: £9100 ÷ 13 = £700 (= 1 part)

3) MULTIPLY TO FIND THE AMOUNTS:
   We want to know Mo's share, which is 4 parts: 4 parts = 4 × £700 = £2800

Watch out for pesky proportional division questions that don't give you the total amount.
You can't just follow the method above, you'll have to be a bit more crafty.

EXAMPLE   **A baguette is cut into 3 pieces. The second piece is twice as long as the first and the third piece is five times as long as the first.**

a) **Find the ratio of the lengths of the 3 pieces.**
   Give your answer in its simplest form.

   If the first piece is 1 part,
   then the second piece is 1 × 2 = 2 parts
   and the third piece is 1 × 5 = 5 parts.
   So the ratio of the lengths = 1:2:5.

b) **The first piece is 28 cm smaller than the third piece.**
   How long is the second piece?

   1) Work out how many parts 28 cm makes up.   28 cm = 3rd piece − 1st piece
       = 5 parts − 1 part = 4 parts

   2) Divide to find one part.   28 cm ÷ 4 = 7 cm

   3) Multiply to find the length of the 2nd piece.   2nd piece = 2 parts = 2 × 7 cm = 14 cm

---

You need to know how to answer all kinds of ratio questions

A good place to start is often to find one part and then go from there. Try that out in some of these questions.

Q1   Orange squash is made of water and concentrate in the ratio 11:2.
      a) What fraction of the squash is made up from concentrate?   [1 mark]
      b) How many litres of water are needed if 6 litres of concentrate are used?   [1 mark]

Q2   The ages of Ben, Graham and Pam are in the ratio 5:3:1.
      Their combined age is 108. How old is Graham?   [2 marks]

Q3   Square A has an area of 36 cm². The areas of square A and square B are
      in the ratio 4:9. What is the side length of square B?   [2 marks]

Q4   In an office, the ratio of people who drink tea to people who drink coffee is 8:5.
      18 more people drink tea than coffee. How many people drink coffee?   [3 marks]

Warm-up and Worked Exam Questions

You'll have to know ratios inside out for your exam — try these quick warm-up questions first and then you'll be ready to tackle those tricky exam practice questions on the next page.

Warm-up Questions

1) Write these ratios in their simplest forms:
   a) 4:8   b) 12:27   c) 1.2:5.4   d) $\frac{8}{3} : \frac{7}{6}$ e) 0.5 litres:400 ml

2) Reduce 5:22 to the form 1:n.

3) A recipe uses flour and sugar in the ratio 3:2.
   How much flour do you need if you're using 300 g of sugar?

4) A nursery group has 12 girls and 6 boys.
   a) Write the ratio of girls to boys.   b) What fraction of the class are girls?

5) Anja collects mugs. The ratio of red mugs to the total number of mugs is 6:15.
   Given that Anja has 50 mugs, how many of them are red?

6) Divide £2400 in the ratio 5:7.

7) Divide 180 in the ratio 3:4:5.

Worked Exam Question

I'm sure you're raring to get stuck into those exam practice questions — before you do, here's one I answered earlier. Read through it carefully and follow the working.

1   Pearl is making a fruit punch. She mixes apple juice,
     pineapple juice and cherryade in the ratio 4:3:7.

   a) What fraction of the fruit punch is pineapple juice? (**3**)

      4 + 3 + 7 = 14 parts in total

      3 of the 14 parts
      are pineapple juice ———— $\frac{3}{14}$
      ....
      [1 mark]

   b) She makes 700 ml of fruit punch. What volume of each drink does she use?

      1 part = 700 ÷ (4 + 3 + 7)   First find how many ml
            = 700 ÷ 14 = 50 ml ————   1 part is equal to.

      Multiply to find the
      amounts of each juice   Apple juice: 50 ml × 4 = 200 ml
      — apple juice is 4 parts,   Pineapple juice: 50 ml × 3 = 150 ml
      pineapple juice is 3 parts   Cherryade: 50 ml × 7 = 350 ml
      and cherryade is 7 parts.

                                    Apple juice: ....200.......... ml

                                    Pineapple juice: ......150........... ml

                                    Cherryade: ....350......... ml
                                                           [3 marks]

Exam Questions

2   The grid on the right shows two shapes, A and B.   ③

Give the following ratios in their simplest form.

a)   Shortest side of shape A : shortest side of shape B

....
[2 marks]

b)   Area of shape A : area of shape B

....
[3 marks]

3   Last month a museum received £21 000 in donations. After taking off the cost of monthly bills, the museum spent the remaining money on new exhibitions.   ③
The ratio of bills to donations was 5 : 14. How much did they spend on new exhibitions?

£ ....
[3 marks]

4   Mr Appleseed's Supercompost is made by mixing soil, compost and grit in the ratio 4 : 3 : 1.
Soil costs £8 per 40 kg, compost costs £15 per 25 kg and grit costs £12 per 15 kg.   ⑤
What is the total cost of materials for 16 kg of Mr Appleseed's Supercompost?

Start by working out how much
of each material is needed for
16 kg of Supercompost.

£ ....
[5 marks]

Direct Proportion Problems

*Direct proportion problems all involve amounts that **increase** or **decrease** together.*

Learn the **Golden Rule** for **Proportion** Questions

There are lots of exam questions which at first sight seem completely different, but in fact they can all be done using the <u>GOLDEN RULE</u>...

Divide for ONE, then Times for ALL

EXAMPLE   **5 pints of milk cost £1.30. How much will 3 pints cost?**

The <u>GOLDEN RULE</u> tells you to:
<u>Divide the price by 5</u> to find how much <u>FOR ONE PINT</u>, then <u>multiply by 3</u> to find how much <u>FOR 3 PINTS</u>.

1 pint: £1.30 ÷ 5 = 0.26 = 26p
3 pints: 26p × 3 = 78p

EXAMPLE   **Emma is handing out some leaflets. She gets paid per leaflet she hands out.
If she hands out 300 leaflets she gets £12.
How many leaflets will she have to hand out to earn £42.50?**

<u>Divide by £12</u> to find how many leaflets she has to hand out to earn <u>£1</u>.

To earn £1: 300 ÷ £12 = 25 leaflets

<u>Multiply by £42.50</u> to find how many leaflets she has to hand out to earn <u>£42.50</u>.

To earn £42.50: 25 × £42.50 = 1062.5
So she'll need to hand out 1063 leaflets.

You need to round your answer up because 1062 wouldn't be enough.

Scaling **Recipes** Up or Down

EXAMPLE   **Judy is making orange and pineapple punch
using the recipe shown on the right.
She wants to make enough to serve 20 people.
How much of each ingredient will Judy need?**

Fruit Punch (serves 8)
800 ml orange juice
140 g fresh pineapple

The <u>GOLDEN RULE</u> tells you to <u>divide each amount by 8</u> to find how much <u>FOR ONE PERSON</u>, then <u>multiply by 20</u> to find how much <u>FOR 20 PEOPLE</u>.

So for 1 person you need:
800 ml ÷ 8 = 100 ml orange juice   ⇒
140 g ÷ 8 = 17.5 g pineapple   ⇒

And for 20 people you need:
20 × 100 ml = 2000 ml orange juice
20 × 17.5 g = 350 g pineapple

Divide for one, then times for all... divide for one, then times for all... divide for one...

Memorise this golden rule — it'll help make direct proportion questions a lot easier. Examiners love recipe questions so learn how to use this golden rule to scale the ingredients up and down. Then try these questions.

Q1   Seven pencils cost £1.40.
a) How much will four pencils cost?   [2 marks]
b) What is the maximum number of pencils you could buy for £6.50?   [2 marks]

Q2   It costs £43.20 for 8 people to go on a rollercoaster 6 times.
How much will it cost for 15 people to go on a rollercoaster 5 times?   [4 marks]

Q2 Video Solution

Direct Proportion Problems

There are many types of direct proportion question — here are another couple for you to learn.

Best Buy Questions

A slightly different type of direct proportion question is comparing the 'value for money' of 2 or 3 similar items. For these, follow the second GOLDEN RULE...

Divide by the PRICE in pence (to get the amount per penny)

EXAMPLE

The local 'Supplies 'n' Vittals' stocks two sizes of Jamaican Gooseberry Jam, as shown on the right. Which of these represents better value for money?

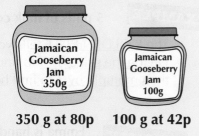

Jamaican Gooseberry Jam 350g   Jamaican Gooseberry Jam 100g

350 g at 80p   100 g at 42p

Follow the GOLDEN RULE —
divide by the price in pence to get the amount per penny.

In the 350 g jar you get   350 g ÷ 80p = 4.38 g per penny
In the 100 g jar you get   100 g ÷ 42p = 2.38 g per penny

The 350 g jar is better value for money, because you get more jam per penny.

In some cases it might be easier to divide by the weight to get the cost per gram.
If you're feeling confident then you can do it this way — if not, the golden rule always works.

Graphing Direct Proportion

Two things are in direct proportion if, when you plot them on a graph, you get a straight line through the origin.

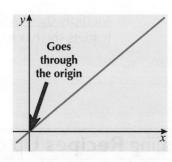

Goes through the origin

Remember, the general equation for a straight line through the origin is $y = Ax$ (see p.61) where A is a number.
All direct proportions can be written as an equation in this form.

EXAMPLE

The amount of petrol, p litres, a car uses is directly proportional to the distance, d km, that the car travels. The car used 12 litres of petrol on a 160 km journey.

a) Write an equation in the form $p = Ad$ to represent this direct proportion.

1) Put the values of $p = 12$ and $d = 160$ into the equation to find the value of A.

$$12 = A \times 160$$
$$A = \frac{12}{160}$$
$$A = 0.075$$

2) Put the value of A back into the equation.

$$p = 0.075d$$

b) Sketch the graph of this direct proportion, marking two points on the line.

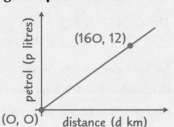

petrol (p litres) | (160, 12)
(0, 0)   distance (d km)

Direct proportion graphs are always straight lines through the origin

Graphing direct proportions is an important skill so put your knowledge to the test with these questions.

Q1   Tomato ketchup comes in bottles of three sizes: 250 g for 50p, 770 g for £1.40 and 1600 g for £3.20. Which bottle represents the best value for money?   [3 marks]

Q2   Brass is made by mixing copper and zinc in the ratio 3:2 by weight.
   a) Sketch a graph showing the weight of copper against the weight of zinc in a sample of brass. Mark two points on the graph.   [3 marks]
   b) What aspects of the graph show that copper and zinc are in direct proportion?   [1 mark]

Inverse Proportion Problems

Here's a trickier type of proportion — but once you've learnt this page you'll be an expert.

Graphing Inverse Proportion

When two things are in inverse proportion, one increases as the other decreases.

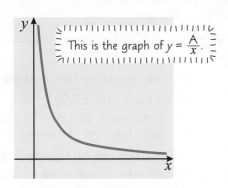

This is the graph of $y = \frac{A}{x}$.

On the graph you can see that as the value of <u>x increases</u>, the value of <u>y decreases</u>. E.g. if x is <u>doubled</u>, y is <u>halved</u>, or if x is <u>multiplied by 5</u>, y is <u>divided by 5</u>.

The general equation for inverse proportion is $y = \frac{A}{x}$.

EXAMPLE Circle each of the equations below that show that *s* is inversely proportional to *t*.

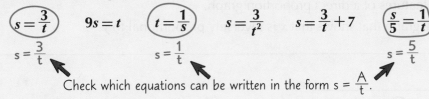

$$\left(s = \frac{3}{t}\right) \qquad 9s = t \qquad \left(t = \frac{1}{s}\right) \qquad s = \frac{3}{t^2} \qquad s = \frac{3}{t} + 7 \qquad \left(\frac{s}{5} = \frac{1}{t}\right)$$

$s = \frac{3}{t}$ $s = \frac{1}{t}$ $s = \frac{5}{t}$

Check which equations can be written in the form $s = \frac{A}{t}$.

Solving Inverse Proportion Questions

On page 83 you saw the 'divide and times' method for direct proportions. Well, inverse proportions are the opposite so you have to:

TIMES for ONE, then DIVIDE for ALL

EXAMPLES

1. It takes 3 farmers 10 hours to plough a field. How long would it take 6 farmers?

<u>Multiply by 3</u> to find how long it would take <u>1 farmer</u>.   10 × 3 = 30 hours for 1 farmer

<u>Divide by 6</u> to find how long it would take <u>6 farmers</u>.   30 ÷ 6 = 5 hours for 6 farmers

Another way of looking at this question is that there are <u>twice</u> as many farmers, so it will take <u>half</u> as long (10 ÷ 2 = 5 hours).

2. 4 bakers can decorate 100 cakes in 5 hours.
If 5 bakers work at the same rate, how much quicker would they decorate 100 cakes?

<u>Multiply by 4</u> to find how long it would take <u>1 baker</u>.   5 hours × 4 = 20 hours for 1 baker

<u>Divide by 5</u> to find how long it would take <u>5 bakers</u>.   20 ÷ 5 = 4 hours for 5 bakers

So 5 bakers are 5 − 4 = 1 hour quicker than 4.

Think of inverse proportions as the opposite of direct proportions

You should be able to identify, draw the graph of and solve inverse proportions. Have a go at solving these:

Q1   It takes 2 carpenters 6 hours to make a bookcase.
How long would it take 8 carpenters to make a bookcase?   [2 marks]

Q2   *m* is inversely proportional to *n*.
When $m = 20$, $n = 5$. What is the value of *m* when *n* is 32?   [2 marks]

Q2 Video Solution

Warm-up and Worked Exam Questions

Question pages take up a large proportion of this section, but that's just because we think they'll be really useful. Give these warm-up questions a shot when you're ready.

Warm-up Questions

1) If three chocolate bars cost 96p, how much will four of the bars cost?

2) 5 red roses cost £7.50.
   a) How much will 8 red roses cost?
   b) Afua has £20. How many red roses can she buy?

3) Marmalade can be bought in 3 different sizes: 250 g (£1.25), 350 g (£2.10) or 525 g (£2.50). Which size is best value for money?

4) It costs £240 to feed 2 elephants for 4 days.
   How much will it cost to feed 1 elephant for 7 days?

5) Give two features of a direct proportion graph.

6) Sketch the graph that shows that x is inversely proportional to y.

Worked Exam Questions

I've worked through one direct proportion and one inverse proportion question for you.
Go through both questions carefully and you'll be ready to tackle the next page.

1   Rani gets paid the same hourly rate whenever she works.
    In the first week of July, Rani worked for 28 hours and got paid £231.
    In each of the next 3 weeks of July, she worked for 25 hours.

    How much will Rani get paid in total for the 4 weeks she worked in July?

    Each hour Rani gets paid: £231 ÷ 28 = £8.25

    In the next three weeks she works 25 × 3 = 75 hours

    So in total she gets paid £231 + (75 × £8.25) = £849.75

    £ ......849.75......
    [2 marks]

2   Circle the **two** equations below that show that f is inversely proportional to g.

$$f = g^2 \qquad f = g + 5 \qquad \boxed{fg = 7} \qquad f = \frac{g}{5} \qquad \boxed{g = \frac{3}{f}}$$

Look for equations which can be rearranged to have the form $f = \frac{A}{g}$, where A is a number.

[2 marks]

Exam Questions

3   Fresh orange juice can be bought in three different sizes.
The price of each is shown on the right.
Which size of bottle is the best value for money?

250 ml £2.00   330 ml £2.75   525 ml £3.75

.... ml
[3 marks]

4   A football coach buys a bottle of water for each child in a football club.
All the bottles of water are the same price. There are 42 boys in the club.
He spends £52.50 on water for the boys. He spends £35 on water for the girls.

How many girls are there in the football club?

....
[2 marks]

5   A ship has enough food to cater for 250 people for 6 days.

a)   For how many days can it cater for 300 people?

.... days
[2 marks]

b)   How many more people can it cater for on a 2-day cruise than on a 6-day cruise?

.... people
[3 marks]

6   Bryn and Richard have just finished playing a game.
The ratio of Bryn's points to Richard's was 5:2.

a) On a set of axes, draw a graph that could be used to work out Bryn's points
if you know Richard's points.

[2 marks]

b) Richard scored 22 points. How many points did Bryn score?

.... points
[1 mark]

Percentages

*You're going to see **6 different types** of percentage question on the next three pages.*
The first few shouldn't give you too much trouble. Especially if you remember:

> 1) 'Per cent' means 'out of 100', so 20% means '20 out of 100' = $\frac{20}{100}$.
>
> 2) To work out the percentage OF something, replace the word OF with a multiplication (×).

Six **Different** Question Types

Type 1 — "Find x% of y"

If you have a calculator, turn the percentage into a decimal, then multiply.

> **EXAMPLE**   **Find 18% of £4.**
>
> Change 18% to a decimal and multiply.
>
> 18% of £4 $\longrightarrow$ Replace 'of' with '×'.
> = 18% × £4
> = 0.18 × £4 = £0.72

If you don't have a calculator, you can use this clever method instead:

> **EXAMPLE**   **Find 135% of 600 kg.**
>
> *You can also find 1% by dividing by 100.*
>
> 100% = 600 kg
>
> 1) Find 10% by dividing by 10:   10% = 600 ÷ 10 = 60 kg
> 2) Find 5% by dividing 10% by 2:   5% = 60 ÷ 2 = 30 kg
> 3) Use these values to make 135%:   135% = 100% + (3 × 10%) + 5%
>   = 600 + (3 × 60) + 30 = 810 kg

Type 2 — "Express x as a percentage of y"

Divide *x* by *y*, then multiply by 100.

> **EXAMPLES**   **1. Give 36p as a percentage of 80p.**
>
> *If you don't have a calculator you'll have to simplify the fraction (see p.15).*
>
> Divide 36p by 80p, then multiply by 100: $\frac{36}{80} \times 100 = 45\%$

> **2. Farmer Littlewood measured the width of his prized pumpkin at the start and end of the month. At the start of the month it was __84 cm__ wide and at the end of the month it was __1.32 m__ wide. Give the __width at the end__ of the month __as a percentage__ of the __width at the start__.**
>
> 1) Make sure both amounts are in the same units.   1.32 m = 132 cm
>
> 2) Divide 132 cm by 84 cm, then multiply by 100: $\frac{132}{84} \times 100 = 157.14\%$ (2 d.p.)

Percentages are one of the most useful things you'll ever learn

Whenever you open a newspaper, see an advert, watch TV or do a maths exam paper you will see percentages. It's really important you get confident with using them — so here is some practise for you.

Q1   a) Without using a calculator, find 36% of 300.   [3 marks]
       b) Use a calculator to find 139% of 505.   [2 marks]

Q2   A full bottle of pearade holds 1.2 litres. After pouring a glass there is 744 ml left
       in the bottle. What percentage of the original amount is left in the bottle?   [2 marks]

Percentages

Type 3 — **New Amount** After a **% Increase or Decrease**

There are two different ways of finding the new amount after a percentage increase or decrease:

1) Find the % then Add or Subtract.

Find the % of the original amount. Add this on to (or subtract from) the original value.

EXAMPLE
A dress has increased in price by 30%.
It originally cost £40. What is the new price of the dress?

1) Find 30% of £40:   30% of £40 = 30% × £40
                                         = 0.3 × 40 = £12
2) It's an increase, so
   add on to the original: £40 + £12 = £52

2) The Multiplier Method

This time, you first need to find the multiplier — the decimal that represents the percentage change.

E.g. 5% increase is 1.05 (= 1 + 0.05)   26% decrease is 0.74 (= 1 − 0.26)

Then you just multiply the original value by the multiplier and voilà — you have the answer.

A % decrease has a multiplier less than 1,
a % increase has a multiplier greater than 1.

EXAMPLE
A hat is reduced in price by 20% in the sales.
It originally cost £12. What is the new price of the hat?

1) Find the multiplier: 20% decrease = 1 − 0.20 = 0.8
2) Multiply the original value by the multiplier: £12 × 0.8 = £9.60

Type 4 — **Simple Interest**

Compound interest is covered on page 92.

Simple interest means a certain percentage of the original amount only is paid at regular intervals (usually once a year). So the amount of interest is the same every time it's paid.

EXAMPLE
Regina invests £380 in an account which pays 3% simple interest each year.
How much interest will she earn in 4 years?

1) Work out the amount of interest earned in one year: 3% = 3 ÷ 100 = 0.03
                                                 3% of £380 = 0.03 × £380
                                                       = £11.40
2) Multiply by 4 to get the total interest for 4 years: 4 × £11.40 = £45.60

Learn how to solve these simple question types

Another two types of percentages for you here — try the methods out on these practice questions.

Q1   A unicorn costing £4000 is reduced by 15% in a sale. What is its new price?   [2 marks]

Q2   Al puts £110 into a bank account that pays 6% simple interest each year. What will
        his bank balance be after 3 years if he doesn't pay in or take out any money?   [3 marks]

Properties of 2D Shapes

*Here's a nice easy page to get you started on **2D shapes**.*

Line Symmetry

This is where you draw one or more <u>MIRROR LINES</u> across a shape and both sides will <u>fold exactly</u> together.

| 2 LINES OF SYMMETRY | 1 LINE OF SYMMETRY | 1 LINE OF SYMMETRY | 3 LINES OF SYMMETRY | NO LINES OF SYMMETRY | 1 LINE OF SYMMETRY |

Rotational Symmetry

This is where you can <u>rotate</u> the shape into different positions that <u>look exactly the same</u>.

Order 1   Order 2   Order 2   Order 3   Order 4

The <u>ORDER OF ROTATIONAL SYMMETRY</u> is the posh way of saying: 'how many different positions look the same'. You should say the Z-shape above has '<u>rotational symmetry of order 2</u>'.

> When a shape has <u>only 1 position</u> you can <u>either</u> say that it has 'rotational symmetry of <u>order 1</u>' or that it has '<u>NO rotational symmetry</u>'.

Regular Polygons

> In an <u>irregular</u> polygon, the sides and angles aren't all equal.

All the <u>sides</u> and <u>angles</u> in a regular polygon are the <u>same</u>.
Learn the names of these <u>regular polygons</u> and how many <u>sides</u> they have.
(An <u>equilateral triangle</u> and a <u>square</u> are both regular polygons — see p.104 and p.105 for their properties.)

REGULAR PENTAGON
<u>5 sides</u>
<u>5 lines</u> of symmetry
Rotational symmetry of <u>order 5</u>

REGULAR HEXAGON
<u>6 sides</u>
<u>6 lines</u> of symmetry
Rotational symmetry of <u>order 6</u>

REGULAR HEPTAGON
<u>7 sides</u>
<u>7 lines</u> of symmetry
Rotational symmetry of <u>order 7</u>

REGULAR OCTAGON
<u>8 sides</u>
<u>8 lines</u> of symmetry
Rotational symmetry of <u>order 8</u>

REGULAR NONAGON
<u>9 sides</u>
<u>9 lines</u> of symmetry
Rotational symmetry of <u>order 9</u>

REGULAR DECAGON
<u>10 sides</u>
<u>10 lines</u> of symmetry
Rotational symmetry of <u>order 10</u>

Make sure you learn the two types of symmetry

They'll get you some nice marks in the exam. Regular polygons have the same number of lines of symmetry and the same order of rotational symmetry as the number of sides.

Q1   Make two copies of the pattern to the right.
      a) Shade two squares to make a pattern with one line of symmetry.   [2 marks]
      b) Shade two squares to make a pattern with rotational symmetry of order 2.   [2 marks]

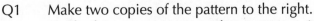

Properties of 2D Shapes

*These two pages have a load of details about **triangles** and **quadrilaterals** — these come in handy for all sorts of geometry problems, so you need to learn them all.*

Triangles

1) **Equilateral** Triangles

3 equal sides and
3 equal angles of 60°.
3 lines of symmetry,
rotational symmetry order 3.

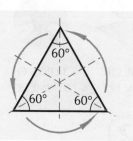

2) **Right-angled** Triangles

1 right angle (90°).
No lines of symmetry.
No rotational symmetry.

The little square means
it's a right angle.

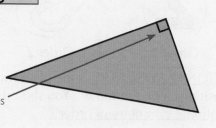

3) **Isosceles** Triangles

2 sides the same.
2 angles the same.
1 line of symmetry.
No rotational symmetry.

These dashes mean
that the two sides are
the same length.

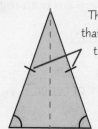

4) **Scalene** Triangles

An acute-angled triangle
has 3 acute angles, and an
obtuse-angled triangle has
one obtuse angle (see p.126).

All three sides different.
All three angles different.
No symmetry (pretty obviously).

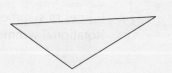

Triangles have three sides

Learn the names (and how to spell them) and the properties of all the triangles on this page.
These are easy marks in the exam — make sure you know them all.

Properties of 2D Shapes

Quadrilaterals

1) Square

 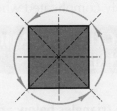

<u>4 equal angles</u> of <u>90°</u> (<u>right angles</u>).
<u>4 lines</u> of symmetry,
rotational symmetry <u>order 4</u>.

2) Rectangle

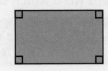

<u>4 equal angles</u> of <u>90°</u> (<u>right angles</u>).
<u>2 lines</u> of symmetry, rotational symmetry <u>order 2</u>.

3) Rhombus

(A square pushed over)
Matching arrows show parallel sides.

A rhombus is the same as a <u>diamond</u>.

 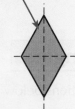

<u>4 equal sides</u> (opposite sides are <u>parallel</u>).
<u>2 pairs</u> of <u>equal angles</u>.
<u>2 lines</u> of symmetry, rotational symmetry <u>order 2</u>.

4) Parallelogram

(A rectangle pushed over)

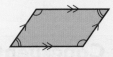 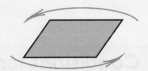

<u>2 pairs</u> of <u>equal sides</u> (each pair are <u>parallel</u>).
<u>2 pairs</u> of <u>equal angles</u>.
<u>NO lines</u> of symmetry,
rotational symmetry <u>order 2</u>.

5) Trapezium

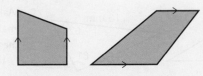

<u>1 pair</u> of <u>parallel sides</u>. <u>NO lines</u> of symmetry*.
No rotational symmetry.

*In an isosceles trapezium, the sloping sides are
the same length. An isosceles trapezium has
1 line of symmetry.

6) Kite

<u>2 pairs</u> of <u>equal sides</u>.
<u>1 pair</u> of <u>equal angles</u>.
<u>1 line</u> of symmetry.
No rotational symmetry.

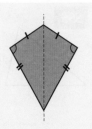

Quadrilaterals have four sides

There are six types of quadrilateral to learn on this page. Make sure you know their features — for example, how many lines of symmetry they have and whether they have rotational symmetry. Now try this question:

Q1   A quadrilateral has all 4 sides the same length and two pairs of equal angles.
        Identify the quadrilateral, and write down its order of rotational symmetry.   [2 marks]

Congruent Shapes

*Shapes can be **congruent**, which basically just means 'the same as each other'.*
Luckily for you, here's a full page on congruence.

Congruent — Same Shape, Same Size

Congruence is another ridiculous maths word which sounds really complicated when it's not:

> ### If two shapes are CONGRUENT, they are EXACTLY THE SAME
> ### — the SAME SIZE and the SAME SHAPE.

These shapes are all congruent:

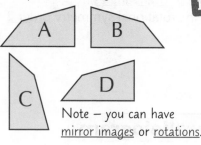

Note — you can have
mirror images or rotations.

EXAMPLE **Two of the triangles below are congruent.**
Write down the letters of the congruent triangles.

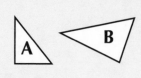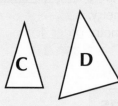

Just pick out the two triangles that are exactly the same —
remember that the shape might have been rotated or reflected.
By eye, you can see that the congruent triangles are **B and E**.

Conditions for Congruent Triangles

Two triangles are congruent if one of the four conditions below holds true:

SSS	three sides are the same
AAS	two angles and a corresponding side match up
SAS	two sides and the angle between them match up
RHS	a right angle, the hypotenuse and one other side all match up

The hypotenuse is the
longest side of a right-angled
triangle — the one opposite
the right angle.

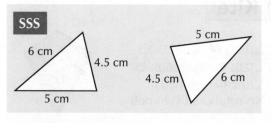

Make sure the sides
match up — here,
the side is opposite
the 81° angle.

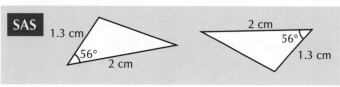

Congruent just means same size, same shape

You need to be able to recognise congruent shapes and learn all 4 conditions for
congruent triangles. Make sure you're using the right sides and angles in each shape.

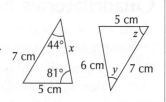

Q1   The two triangles shown on the right are congruent.
Find the values of *x*, *y* and *z*.   [2 marks]

Similar Shapes

Similar shapes are **exactly the same shape**, but can be **different sizes** (they can also be **rotated** or **reflected**).

SIMILAR — same shape, **different size**.

Similar Shapes Have the Same Angles

Generally, for two shapes to be <u>similar</u>, all the <u>angles</u> must match and the <u>sides</u> must be <u>proportional</u>. But for <u>triangles</u>, there are <u>three special conditions</u> — if any one of these is true, you know they're similar.

Two triangles are similar if:

1) All the <u>angles</u> match up.

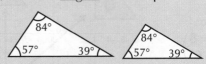

2) All three <u>sides</u> are <u>proportional</u>.

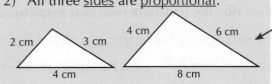

Here, the sides of the bigger triangle are twice as long as the sides of the smaller triangle.

3) Any <u>two sides</u> are <u>proportional</u> and the <u>angle between them</u> is the <u>same</u>.

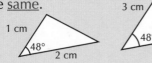

Watch out — if one of the triangles has been rotated or flipped over, it might look as if they're not similar, but don't be fooled.

EXAMPLE

Tony says, "Triangles ABC and DEF are similar." Is Tony correct? Explain your answer.

<u>Check</u> condition 3 holds — start by finding the <u>missing angle</u> in triangle DEF:

Angle DEF = 180° − 46° − 30° = 104° so angle ABC = angle DEF

Now check that <u>AB</u> and <u>BC</u> are <u>proportional</u> to <u>DE</u> and <u>EF</u>:

DE ÷ AB = 6 ÷ 2 = 3 and EF ÷ BC = 9 ÷ 3 = 3 so DE and EF are 3 times as long as AB and BC.

Tony is correct — two sides are proportional and the angle between them is the same so the triangles are similar.

Use Similarity to Find Missing Lengths

You might have to use the <u>properties</u> of similar shapes to find missing distances, lengths etc. — you'll need to use <u>scale factors</u> (see p.109) to find the lengths of missing sides.

EXAMPLE

ABC and ADE are similar right-angled triangles. AC = 20 cm, AE = 50 cm and BC = 8 cm. Find the length of DE.

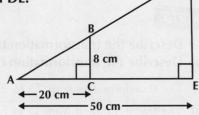

The triangles are <u>similar</u>, so work out the <u>scale factor</u>:

scale factor = $\frac{50}{20}$ = <u>2.5</u>

Now <u>use</u> the scale factor to work out the length of DE:

DE = 8 × 2.5 = 20 cm

Similar means the same shape but a different size

To help remember the difference between similarity and congruence, think '<u>similar siblings, congruent clones</u>' — siblings are alike but not the same, clones are identical.

Q1   Find the length of DB.   [3 marks]

Q1 Video Solution

The Four Transformations

*There are four **transformations** you need to know — **translation**, **rotation**, **reflection** and **enlargement**.*

1) Translations

In a <u>translation</u>, the <u>amount</u> the shape moves by is given as a <u>vector</u> (see p.147-148) written $\begin{pmatrix} x \\ y \end{pmatrix}$ — where x is the <u>horizontal movement</u> (i.e. to the <u>right</u>) and y is the <u>vertical movement</u> (i.e. up). If the shape moves <u>left and down</u>, x and y will be <u>negative</u>.

EXAMPLE   a) **Describe the transformation that maps triangle P onto Q.**
b) **Describe the transformation that maps triangle P onto R.**

a) To get from P to Q you need to move <u>8 units left</u> and <u>6 units up</u>, so...
The transformation from P to Q is a translation by the vector $\begin{pmatrix} -8 \\ 6 \end{pmatrix}$.

b) The transformation from P to R is a translation by the vector $\begin{pmatrix} 0 \\ 7 \end{pmatrix}$.

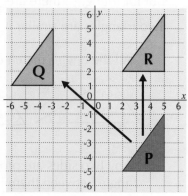

2) Rotations

To describe a <u>rotation</u>, you must give <u>3 details</u>:
1) The <u>angle of rotation</u> (usually 90° or 180°).
2) The <u>direction of rotation</u> (clockwise or anticlockwise). ← *For a rotation of 180°, it doesn't matter whether you go clockwise or anticlockwise.*
3) The <u>centre of rotation</u> (often, but not always, the origin).

EXAMPLE   a) **Describe the transformation that maps triangle A onto B.**
b) **Describe the transformation that maps triangle A onto C.**

a) The transformation from A to B is a rotation of <u>90°</u> <u>anticlockwise</u> about the <u>origin</u>.

b) The transformation from A to C is a rotation of <u>180°</u> clockwise (or anticlockwise) about the <u>origin</u>.

If it helps, you can use tracing paper to help you find the centre of rotation.

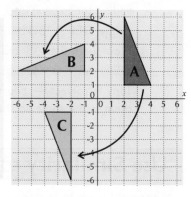

3) Reflections

For a <u>reflection</u>, you must give the <u>equation</u> of the <u>mirror line</u>.

EXAMPLE

a) **Describe the transformation that maps shape D onto shape E.**
b) **Describe the transformation that maps shape D onto shape F.**

a) The transformation from D to E is a reflection in the y-axis.
b) The transformation from D to F is a reflection in the line $y = x$.

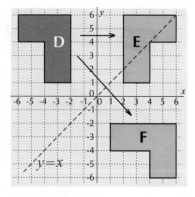

A rotation is specified by an angle, a direction and a centre

Shapes are <u>congruent</u> under translation, reflection and rotation — this is because their <u>size</u> and <u>shape</u> don't change, just their position and orientation. Now have a go at this Exam Practice Question:

Q1   On a grid, copy shape D above and rotate it 90° clockwise about the point (−1, −1).   [2 marks]

The Four Transformations

*One more transformation coming up — **enlargements**. They're the trickiest, but also the most interesting.*

4) Enlargements

For an enlargement, you must specify:

1) The scale factor. ←
2) The centre of enlargement.

$$\text{scale factor} = \frac{\text{new length}}{\text{old length}}$$

1) The scale factor for an enlargement tells you how long the sides of the new shape are compared to the old shape. E.g. a scale factor of 3 means you multiply each side length by 3.

2) If you're given the centre of enlargement, then it's vitally important where your new shape is on the grid.

> The scale factor tells you the **RELATIVE DISTANCE** of the
> old points and new points from the centre of enlargement.

So, a scale factor of 2 means the corners of the enlarged shape are twice as far from the centre of enlargement as the corners of the original shape.

Describing Enlargements

EXAMPLE

Describe the transformation that maps Triangle A onto Triangle B.

Use the formula to find the scale factor. (Just do this for one pair of sides.)

Old length of triangle base = 3 units
New length of triangle base = 6 units

Scale factor = $\frac{\text{new length}}{\text{old length}} = \frac{6}{3} = 2$

To find the centre of enlargement, draw lines that go through matching corners of both shapes and see where they cross.

So the transformation is an enlargement of scale factor 2, centre (2, 6).

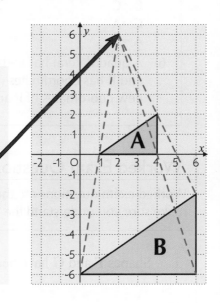

Fractional Scale Factors

1) If the scale factor is bigger than 1 the shape gets bigger.
2) If the scale factor is smaller than 1 (e.g. ½) it gets smaller.

EXAMPLE

Enlarge the shaded shape by a scale factor of $\frac{1}{2}$, about centre 0.

1) Draw lines going from the centre to each corner of the original shape. The corners of the new shape will be on these lines.
2) The scale factor is $\frac{1}{2}$, so make each corner of the new shape half as far from 0 as it is in the original shape.

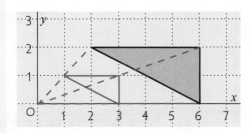

An enlargement is given by a scale factor and a centre of enlargement

Shapes are similar under enlargement — the position and the size change, but the angles and ratios of the sides don't (see p.107). A scale factor smaller than 1 means the shape gets smaller.

Q1 On a grid, draw triangle A with vertices (2, 1), (4, 1) and (4, 3), and triangle B with vertices (3, 1), (7, 1) and (7, 5). Describe the transformation that maps A to B. [4 marks]

Q1 Video Solution

Warm-up and Worked Exam Questions

Nothing too tricky so far in this section. Now it's time for some warm-up questions to get your brain ticking — before moving on to the exam-style questions. It's all good practice for the big day.

Warm-up Questions

1) Copy the letters, mark in their lines of symmetry and give their order of rotational symmetry:

C W I D Q

2) Give all the properties of an equilateral triangle.

3) How many lines of symmetry does a kite have?

4) Which two of these shapes are: a) similar?   b) congruent?

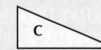

5) Describe fully these 4 transformations, as shown on the grid:
   A→B
   B→C
   C→A
   A→D

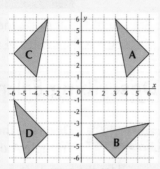

6) A quadrilateral has vertices PQRS, where P has coordinates (3, 4) and R has coordinates (5, −1). The quadrilateral is translated, and the new coordinates of P are (7, 2). What are the new coordinates of R?

Worked Exam Question

Worked exam questions are the ideal way to get the hang of answering the real exam questions — make sure you understand the answer to this one.

1   Shapes **F** and **G** have been drawn on the grid below. **(3)**

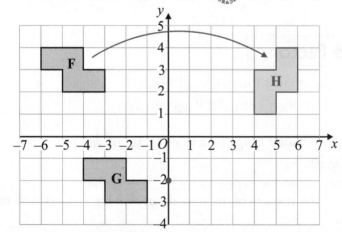

a)   Write down the vector which describes the translation that maps **F** onto **G**.

To get from F to G, the shape is moved 2 units to the right and 5 units down.

$$\begin{pmatrix} 2 \\ -5 \end{pmatrix}$$
....
[2 marks]

b)   Rotate shape **F** by 90° clockwise about the point (0, −2). Label your image **H**.

See the grid above — you might find it easiest to use tracing paper to draw rotations.

[2 marks]

Exam Questions

2   An isosceles triangle has vertices A(1, 1), B(3, 7) and C(5, 1). ②
     Give the equation of its line of symmetry.

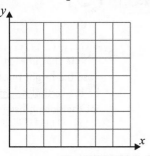

....
                                                                    [1 mark]

3   Triangle **A** has been drawn on the grid below. ③
     Reflect triangle **A** in the line $x = -1$. Label your image **B**.

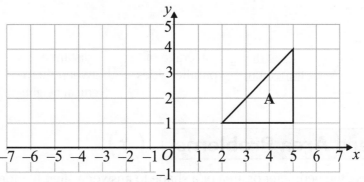

                                                                    [2 marks]

4   The shapes *ABCD* and *EFGH* are mathematically similar. ④

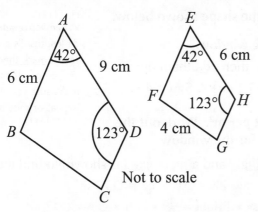   a)   Find the length of *EF*.

                              .... cm
                                                 [2 marks]

     b)   Find the length of *BC*.

Not to scale

                              .... cm
                                                 [1 mark]

5   On the grid enlarge the triangle by a scale factor of 3, centre (–4, 0). ④

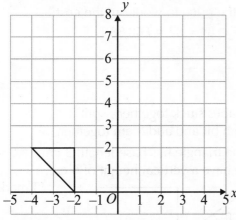

                                                                    [3 marks]

Perimeter and Area

Perimeter is the **distance** around the outside of a shape. **Area** is a bit trickier — you need to learn some *formulas*. You should already know that the area of a **rectangle** is $A = l \times w$ and the area of a **square** is $A = l^2$.

Area Formulas for **Triangles** and **Quadrilaterals**

Learn these formulas:

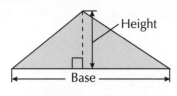

Area of triangle = ½ × base × vertical height

$$A = \tfrac{1}{2} \times b \times h$$

Note that in each case the height must be the <u>vertical height</u>, not the sloping height.

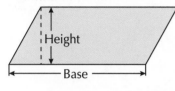

$\dfrac{\text{Area of}}{\text{parallelogram}}$ = base × vertical height

$$A = b \times h$$

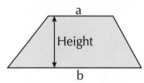

$\dfrac{\text{Area of}}{\text{trapezium}}$ = average of parallel sides × distance between them (vertical height)

$$A = \tfrac{1}{2}(a + b) \times h$$

Perimeter and **Area** Problems

You might have to <u>use</u> the perimeter or area of a shape to answer a <u>slightly more complicated</u> question (e.g. find the area of a wall, then work out how many rolls of wallpaper you need to wallpaper it).

EXAMPLE **Huda is making a stained-glass window in the shape shown below.**

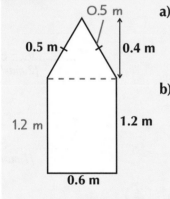

a) **Find the perimeter of the window.**

<u>Label</u> all the side lengths, then <u>add</u> them up:
0.5 m + 1.2 m + 0.6 m + 1.2 m + 0.5 m = 4 m

When you're adding side lengths it's a good idea to <u>mark them off</u> as you go along to make sure you don't repeat or miss any.

b) **Coloured glass costs £82 per m². Work out the cost of the glass needed for the window.**

Split the shape into a <u>triangle</u> and a <u>rectangle</u> (as shown) to find the area:
Area of rectangle = length × width = 0.6 × 1.2 = 0.72 m²
Area of triangle = $\frac{1}{2}$ × base × height = $\frac{1}{2}$ × 0.6 × 0.4 = 0.12 m²
Total area of shape = 0.72 + 0.12 = <u>0.84 m²</u>

Then <u>multiply</u> the <u>area</u> by the <u>price</u> to work out the cost:
Cost = area × price per m² = 0.84 × 82 = £68.88

Learn the area formulas

If you have a composite shape (a shape made up of different shapes stuck together), split it into triangles and quadrilaterals. Then, work out the area of each bit and add them together.

Q2 Video Solution

Q1   A shape is made up of a triangle and a parallelogram. The triangle has base length 3 cm and the parallelogram has base length 11 cm. They both have a vertical height of 6 cm. Find the total area of the shape.

[3 marks]

Q2   The triangle and rectangle shown on the right have the same area. Find the value of x.

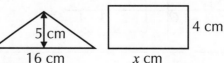

[2 marks]

Perimeter and Area — Circles

Another page of formulas, this time on circles. You know the drill — learn all the formulas in boxes on this page.

Area and Circumference of Circles

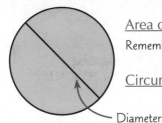

Diameter

<u>Area of circle</u> = π × (radius)^2
Remember that the <u>radius</u> is <u>half</u> the <u>diameter</u>.

<u>Circumference</u> = π × diameter
= 2 × π × radius

$$A = \pi r^2$$

$$C = \pi D = 2\pi r$$

For these formulas, use the π button on your calculator. For non-calculator questions, use $\pi \approx 3.142$.

Arc Lengths and Areas of Sectors

These next ones are a bit more tricky — before you try and <u>learn</u> the <u>formulas</u>, make sure you know what a <u>sector</u> and an <u>arc</u> are (I've helpfully labelled the diagram below — I'm nice like that).

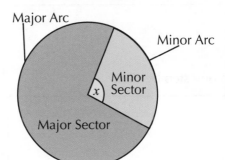

Major Arc

Minor Arc

Minor Sector

Major Sector

x

<u>Area of Sector</u> = $\dfrac{x}{360}$ × Area of full Circle

<u>Length of Arc</u> = $\dfrac{x}{360}$ × Circumference of full Circle

You also need to know what a <u>segment</u>, a <u>chord</u> and a <u>tangent</u> are.

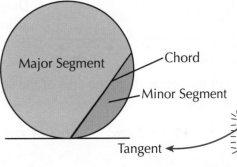

Major Segment

Chord

Minor Segment

Tangent

A tangent <u>just</u> touches <u>one point</u> of the circle.

EXAMPLE

In the diagram on the right, a sector with angle 60° has been cut out of a circle with radius 3 cm. Find the exact area of the shaded shape.

3 cm

300°

Use the formula to find the area of the shaded sector:

area of sector = $\dfrac{x}{360} \times \pi r^2 = \dfrac{300}{360} \times \pi \times 3^2$

$= \dfrac{5}{6} \times \pi \times 9 = \dfrac{15}{2}\pi$ cm^2

'Exact area' means leave your answer in terms of π.

Make sure you know all of these circle terms

One more thing — if you're asked to find the perimeter of a semicircle or quarter circle, don't forget to add on the straight edges too. It's an easy mistake to make, and it'll cost you marks. Now, have a go at this question:

Q1　a) For sector A, find:
　　　　(i)　the area. Give your answer in terms of π.　　[2 marks]
　　　　(ii)　the arc length, to 2 d.p.　　[2 marks]
　　b) Show that area of sector A is the
　　　　same as the area of sector B.　　[2 marks]

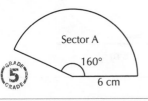

Sector A

Sector B

160°

6 cm

8 cm

Warm-up and Worked Exam Questions

There are lots of formulas in this section. The best way to find out what you know is to practise these questions. If you find you keep forgetting the formulas, you need more practice.

Warm-up Questions

1) Find the perimeter of the shape shown on the right.

2) Give the formulas for:
   a) the area of a rectangle
   b) the circumference of a circle
   c) the area of a parallelogram

3) A triangle has a base of 3 m and a vertical height of 7 m. Calculate its area.

4) A woodworking template has the shape shown on the right: Calculate the area of the template.

5) a) What is a tangent?
   b) Draw a circle and label a tangent.

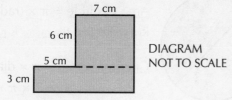

7 cm
6 cm
5 cm
3 cm

DIAGRAM NOT TO SCALE

8 cm

Worked Exam Questions

Here are two juicy worked exam questions for you. Work through each one step by step.

1   The diagram below shows a rectangle and a square. (4)

 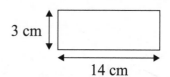

3 cm

14 cm

? cm

Diagram not accurately drawn

The ratio of the area of the rectangle to the area of the square is 6 : 7.
What is the area of the square?

Area of rectangle = 3 × 14 = 42 cm²

See p.79 for more
on scaling up ratios.

area of rectangle : area of square
                    6 : 7
(×7)   42 : 49   (×7)
So area of the square = 49 cm²

.........**49**....... cm²
[2 marks]

2   The diagram below shows a square with a circle inside. The circle touches each of the four sides of the square. Calculate the shaded area. Give your answer to 2 d.p. (4)

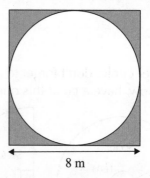

8 m

Area of square = 8 × 8 = 64 m²
Area of circle = π × 4² = 50.2654... m²
Shaded area = 64 − 50.2654...
            = 13.7345... m²

.........**13.73**....... m²
[3 marks]

Exam Questions

3   A shape is made up from an isosceles triangle and a trapezium.

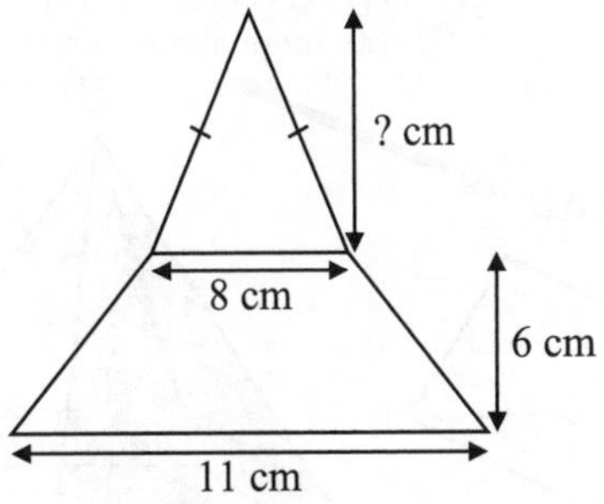

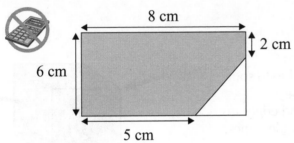

? cm

8 cm

6 cm

11 cm

Diagram not
accurately drawn

The area of the trapezium is 3 times as big as
the area of the triangle.

a)   Find the total area of the shape.

.... cm²
[3 marks]

b)   Find the height of the triangle.

.... cm
[2 marks]

4   The diagram below shows a rectangle with a right-angled triangle inside. **(4)**

8 cm

2 cm

6 cm

5 cm

Calculate the area of the shaded part.

.... cm²
[4 marks]

5   Look at the sector shown in the diagram below. **(5)**

Diagram not
accurately drawn

30°

6 cm

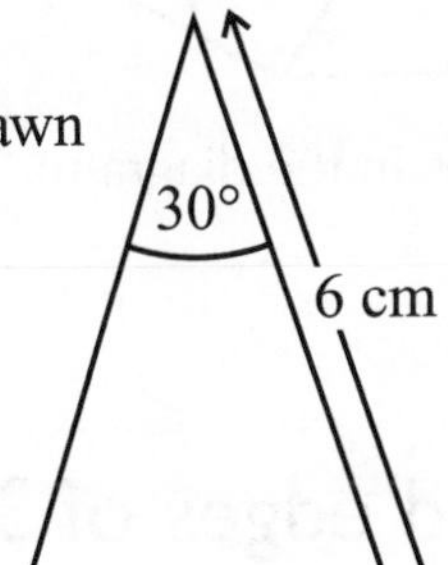

Find the perimeter and the area of the sector.
Give your answers to 3 significant figures.

Don't forget to add
the two radii to the
arc length when
finding the perimeter.

Perimeter = ... cm

Area = ... cm²
[5 marks]

3D Shapes

*First up are some **3D shapes** for you to learn, closely followed by a look at the **different parts of solids**.*

Eight **Solids** to Learn

<u>3D shapes</u> are <u>solid shapes</u>. These are the ones you need to know:

> ⌐ There's more about prisms on p.118. ⌐

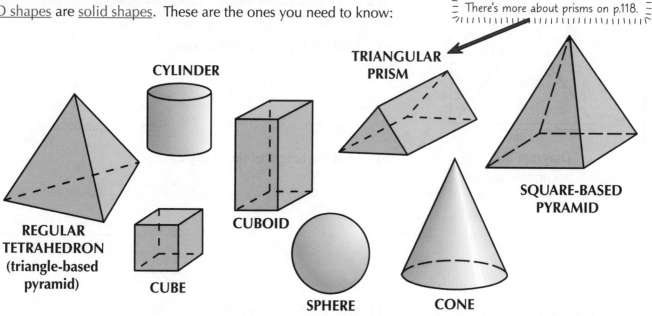

CYLINDER

TRIANGULAR PRISM

SQUARE-BASED PYRAMID

REGULAR TETRAHEDRON (triangle-based pyramid)

CUBOID

CUBE

SPHERE

CONE

Different **Parts** of Solids

There are different parts of 3D shapes you need to be able to spot. These are <u>vertices</u> (corners), <u>faces</u> (the flat bits) and <u>edges</u>. You might be asked for the <u>number</u> of vertices, faces and edges in the exam — just <u>count</u> them up, and don't forget the <u>hidden</u> ones.

> ⌐ Faces (especially curved faces) are sometimes called surfaces. ⌐

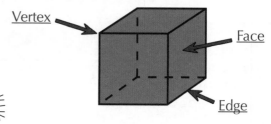

Vertex

Face

Edge

EXAMPLE

For the triangular prism on the right, write down the number of faces, the number of edges and the number of vertices.

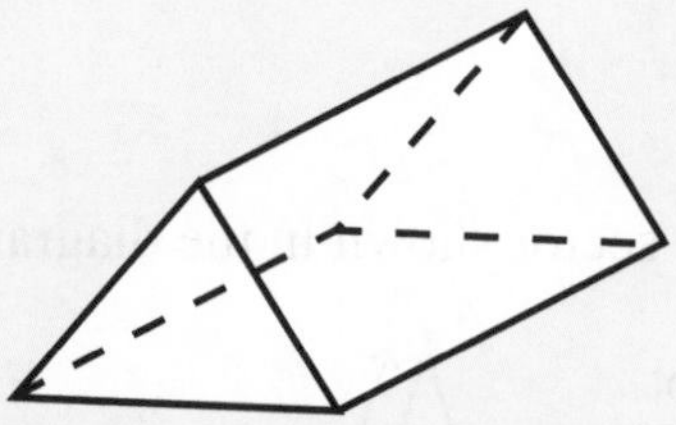

A triangular prism has **5 faces** (there are three rectangular faces and two triangular ones).

It has **9 edges** (there are 3 hidden ones — the dotted lines in the diagram).

It has **6 vertices** (there's one hidden at the back).

Make sure you can spot the vertices, faces and edges of 3D shapes

Remember — 1 vertex, 2 vertices. They're funny words, designed to confuse you, so don't let them catch you out. You need to know the names of all the solid shapes above too.

Q1   a) Write down the mathematical name of the shape on the right.   **[1 mark]**

        b) Write down: (i)   the number of faces   **[1 mark]**

                         (ii)   the number of edges   **[1 mark]**

                         (iii) the number of vertices   **[1 mark]**

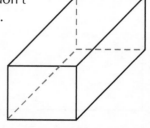

3D Shapes — Surface Area

*Surface area is like normal area but for 3D shapes — for some shapes you can use **2D areas** and add them up, for others there are special **formulas** you'll need to use.*

Surface Area using Nets

1) <u>SURFACE AREA</u> only applies to 3D objects — it's just the <u>total area</u> of all the <u>faces</u> added together.

2) <u>SURFACE AREA OF A SOLID = AREA OF THE NET</u> (remember that a <u>net</u> is just a <u>3D shape</u> folded out flat). So if it helps, imagine (or sketch) the net and add up the area of <u>each bit</u>.

EXAMPLE **Find the surface area of the square-based pyramid below.**

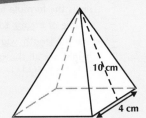

Sketch the net — a square-based pyramid has <u>1 square face</u> and <u>4 triangular faces</u>.
Area of square face = 4 × 4 = 16 cm^2
Area of triangular face = ½ × 4 × 10 = 20 cm^2
Total surface area = 16 + (4 × 20) = 16 + 80 = 96 cm^2

Surface Area Formulas

1) <u>SPHERES, CONES AND CYLINDERS</u> have surface area formulas that you need to be able to use.

2) Luckily you <u>don't</u> need to memorise the <u>sphere</u> and <u>cone</u> formulas — you'll be given them in your exam.

3) But you must get <u>lots of practice</u> using them, or you might slip up when it comes to the exam.

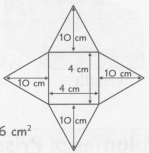

Surface area of a SPHERE = $4\pi r^2$

curved area of cone (l is the slant height)   area of circular base

Surface area of a CONE = $\pi rl + \pi r^2$

Surface area of a CYLINDER = $2\pi rh + 2\pi r^2$

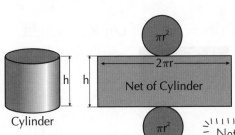

Note that the <u>length of the rectangle</u> is equal to the <u>circumference</u> of the circular ends.

EXAMPLE **Find the surface area of the cylinder on the right to 1 d.p.**

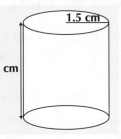

Just put the <u>measurements</u> into the <u>formula</u> and work it out very carefully in stages:
Surface area of cylinder = $2\pi rh + 2\pi r^2$
= (2 × π × 1.5 × 5) + (2 × π × 1.5^2)
= 47.123... + 14.137... = 61.261... = 61.3 cm^2

To find the surface area of a solid, add up the area of each face

Use nets to find surface area — they make it easier to see each face and work out the area.
Once you're familiar with the formulas on this page, get some practice with these questions.

Q2 Video Solution

Q1   Find the surface area of the cone on the right.   [3 marks]   20 cm

Q2   Find the radius of a sphere with surface area of 196π cm^2.   [2 marks]   5 cm

3D Shapes — Volume

*You've already had a couple of pages on 3D shapes — now it's time to work out their **volumes**.*

LEARN these volume formulas...

Volumes of **Cuboids**

A <u>cuboid</u> is a <u>rectangular block</u>. Finding its volume is dead easy:

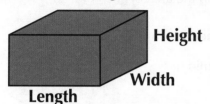

Volume of Cuboid = length × width × height

$$V = L \times W \times H$$

This is the formula for the volume of a <u>cube</u> too — where the <u>length</u>, <u>width</u> and <u>height</u> are all the <u>same</u>.

Volumes of **Prisms**

A <u>PRISM</u> is a solid (3D) object which is the <u>same shape</u> all the way through
— i.e. it has a <u>CONSTANT AREA OF CROSS-SECTION</u>.

Triangular Prism

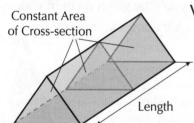

Constant Area of Cross-section

Length

Volume of Prism = cross-sectional area × length

$$V = A \times L$$

This formula works for <u>any</u> prism.

Cylinder
(circular prism)

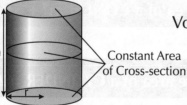

h

Constant Area of Cross-section

r

Volume of Cylinder = area of circle × height

$$V = \pi r^2 h$$

EXAMPLE **Honey comes in cylindrical jars with radius 4.5 cm and height 12 cm.**
Dan has a recipe that needs 1 litre of honey. How many jars should he buy?

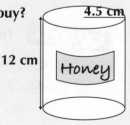

4.5 cm

12 cm

Honey

First, work out the <u>volume</u> of the jar — just use the <u>formula</u> above:

$V = \pi r^2 h = \pi \times 4.5^2 \times 12$
    $= 763.4070... \text{ cm}^3$

1 litre = 1000 cm³ (see p.95), so he needs to buy **2 jars of honey**.

You have to remember what a prism is

You might get a question where you're given a shape made up of 1 cm cubes
and asked for its volume. All you have to do here is count up the cubes
(not forgetting any hidden ones at the back of the shape). Have a go at this
Exam Practice Question — you'll need your area formulas from p.112.

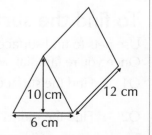

10 cm   12 cm   6 cm

Q1   Find the volume of the triangular prism on the right.   [3 marks]

3D Shapes — Volume

Another page on volumes now — my generosity knows no limits.

Volumes of **Spheres**

$$\text{Volume of Sphere} = \frac{4}{3}\pi r^3$$

A <u>hemisphere</u> is half a sphere. So the volume of a hemisphere is just half the volume of a full sphere, $V = \frac{2}{3}\pi r^3$.

Volumes of **Pyramids** and **Cones**

A pyramid is a shape that goes from a <u>flat base</u> up to a <u>point</u> at the top. Its base can be any shape at all.

Cone   Square-based Pyramid

$$\text{Volume of Pyramid} = \frac{1}{3} \times \text{Base Area} \times \text{Vertical Height}$$
$$\text{Volume of Cone} = \frac{1}{3} \times \pi r^2 \times h$$

Make sure you use the <u>vertical height</u> in these formulas — don't get confused with the <u>slant height</u>, which you used to find the <u>surface area</u> of a cone.

Volumes of **Frustums**

A <u>frustum of a cone</u> is what's left when the top part of a cone is cut off parallel to its circular base. You'll be given the formula for the <u>volume</u> of a <u>cone</u> in your exam, but you'll need to remember:

$$\frac{\text{VOLUME OF}}{\text{FRUSTUM}} = \frac{\text{VOLUME OF}}{\text{ORIGINAL CONE}} - \frac{\text{VOLUME OF}}{\text{REMOVED CONE}}$$

This bit is the frustum

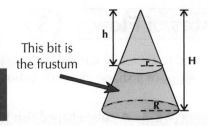

EXAMPLE

A waste paper basket is the shape of a frustum formed by removing a cone of height 10 cm from a cone of height 50 cm and radius 35 cm. Find the volume of the waste paper basket to 3 significant figures.

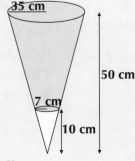

Use the <u>formula</u> for the <u>volume of a cone</u> above:

Volume of original cone $= \frac{1}{3} \times \pi \times 35^2 \times 50 = 64140.850...$ cm^3

Volume of removed cone $= \frac{1}{3} \times \pi \times 7^2 \times 10 = 513.126...$ cm^3

Volume of frustum $= 64140.850... - 513.126... = 63627.723... = 63\,600$ cm^3 (3 s.f.)

Remember, a frustum is just a cone with the top chopped off

The frustum formula makes sense when you think about it — if you subtract the volume of the removed cone from the original volume, you're left with the volume of the bit at the bottom. Try it on this question:

Q1   A cone has radius 12 m and vertical height 18 m. A cone 3 m high with a radius of 2 m is cut off the cone to make a frustum. Find the volume of the frustum, leaving your answer in terms of π.   [4 marks]

Q1 Video Solution

Trickier Volume Problems

*You might get a question asking you something a little more challenging than just finding the volume — this page will talk you through a couple of **trickier types** of volume question.*

Ratios of Volumes

1) You might need to look at how the <u>volumes</u> of different shapes are linked.

2) So you could be given two shapes and have to show <u>how many times bigger</u> one volume is than the other, or you might need to show the <u>ratio</u> of their volumes:

> 1) Work out each volume <u>separately</u> — make sure the <u>units</u> are the <u>same</u>.
> 2) Write the volumes as a <u>ratio</u> and <u>simplify</u>.

 EXAMPLE
The cone in the diagram has a radius of 5 cm and a height of 12 cm. The sphere has a radius of 15 cm. Find the ratio of their volumes in its simplest form.

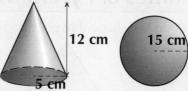

Put the numbers into the <u>volume formulas</u> (see p.119):

Volume of cone $= \frac{1}{3} \times \pi r^2 \times h = \frac{1}{3} \times \pi \times 5^2 \times 12 = \frac{1}{3} \times \pi \times 300 = 100\pi$ cm^3

Volume of sphere $= \frac{4}{3}\pi r^3 = \frac{4}{3} \times \pi \times 15^3 = \frac{4 \times 3375}{3} \times \pi = 4500\pi$ cm^3

> When comparing volumes, it's usually best to leave your answers in terms of π.

Find the <u>ratio</u> of the volumes and <u>simplify</u>: volume of cone:volume of sphere

$\div 100\pi \left(\frac{100\pi : 4500\pi}{1 : 45} \right) \div 100\pi$

Rates of Flow

You need to be really careful with <u>units</u> in rates of flow questions. You might be given the <u>dimensions</u> of a shape in <u>cm</u> or <u>m</u> but the <u>rate of flow</u> in <u>litres</u> (e.g. litres per minute).

 EXAMPLE
A cube-shaped fish tank with sides of length 30 cm is being filled with water at a rate of 4 litres per minute. How long will it take to fill the fish tank? Give your answer in minutes and seconds. 1 litre = 1000 cm^3

Find the <u>volume</u> of the fish tank: V = (side length)^3 = 30^3 = 27 000 cm^3
Then convert the <u>rate of flow</u> into cm^3/minute — <u>multiply</u> by 1000:
4 litres per minute = 4 × 1000 = 4000 cm^3/min
So it will take 27 000 ÷ 4000 = 6.75 minutes
= 6 minutes and 45 seconds to fill the fish tank.

Leave volumes in terms of π when comparing them

Sorry, this page is a rotten one. When dealing with rates of flow, you might need to convert the rate of flow so its units match the volume (e.g. m^3/s), then use this to find your answer.

Q1 A sphere has a radius of 3 cm. A cylinder has a radius of 6 cm and a height of 10 cm. Show that the volume of the cylinder is 10 times as big as the volume of the sphere. [3 marks]

Q2 A square-based pyramid with base sides of length 60 cm and height 110 cm is being filled with water at a rate of 100 cm^3 per second. Does it take longer than 20 minutes to fill? [3 marks]

Projections

Projections are just different **views** of a 3D solid shape — looking at it from the **front**, the **side** and the **top**.

The Three Different **Projections**

There are three different types of projections — <u>front elevations</u>, <u>side elevations</u> and <u>plans</u> (elevation is just another word for projection).

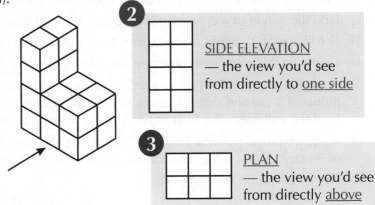

1 <u>FRONT ELEVATION</u> — the view you'd see from directly <u>in front</u> (in the direction of the arrow)

2 <u>SIDE ELEVATION</u> — the view you'd see from directly to <u>one side</u>

3 <u>PLAN</u> — the view you'd see from directly <u>above</u>

Don't be thrown if you're given a diagram on <u>isometric</u> (dotty) paper like this — it works in just the same way. If you have to <u>draw</u> shapes on isometric paper, just <u>join the dots</u>. You should <u>only</u> draw <u>vertical</u> and <u>diagonal lines</u> (no horizontal lines).

Drawing Projections

EXAMPLES

1. The front elevation and plan view of a shape are shown below. Sketch the solid shape.

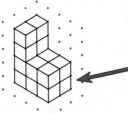

Front Elevation   **Plan View**

Just piece together the original shape from the information given — here you get a <u>prism</u> in the shape of the <u>front elevation</u>.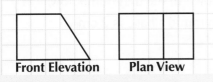

2. a) On the cm square grid, draw the side elevation of the prism from the direction of the arrow.
b) Draw a plan of the prism on the grid.

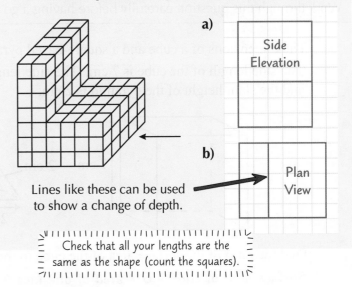

a) Side Elevation

b) Plan View

Lines like these can be used to show a change of depth.

Check that all your lengths are the same as the shape (count the squares).

You need to know the three different types of projection

Projection questions aren't too bad — just take your time and sketch the diagrams carefully. Watch out for questions on isometric paper — they may look confusing, but can actually be easier than other questions.

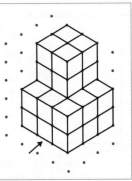

Q1   For the shape on the right, draw:
    a) The front elevation (from the direction of the arrow),   [1 mark]
    b) The side elevation,   [1 mark]
    c) The plan view.   [1 mark]

Warm-up and Worked Exam Questions

Turns out there's lots to know about 3D shapes. If you have any problems doing these warm-up questions, look back over anything you're unsure of before tackling the exam questions.

Warm-up Questions

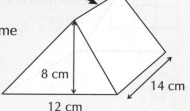

1) How many faces, vertices and edges does a triangle-based pyramid have?
2) Find the height of a cylinder with radius 4 cm and surface area 64π cm².
3) The shape on the right is made from 1 cm cubes. Find its volume.
4) Calculate the volume of the triangular prism below on the right.
5) A cone is 24 cm tall, and has a radius of 9 cm. A cone with a radius of 3 cm and vertical height of 8 cm is removed to make a frustum. Find the volume of the frustum. Give your answer in terms of π.
6) A square-based pyramid has a vertical height of 8 cm, and a volume of 96 cm³. What is the length of each side of the square base?
7) For the shape on the right, draw:
   a) the front elevation (from the direction of the arrow)
   b) the side elevation
   c) the plan view.

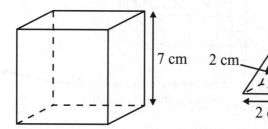

8 cm
14 cm
12 cm

Worked Exam Question

Work through this question carefully before having a go at the exam questions.

1   The dimensions of a cube and a square-based pyramid are shown in the diagram below. **(4)**
    The side length of the cube is 7 cm. The side length of the pyramid's base is 2 cm
    and the slant height of the pyramid is 2 cm.

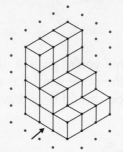

7 cm   2 cm   Not drawn accurately

2 cm

Find the ratio of the surface area of the cube to the surface area of the pyramid in the form $n:1$.

Surface area of cube = 6 × area of one face = 6 × 7 × 7 = 294 cm²

Surface area of square-based pyramid = area of base + (4 × area of one triangular face)
= 2 × 2 + (4 × ½ × 2 × 2) = 4 + 8 = 12 cm²

Surface area of cube : surface area of pyramid
= 294 : 12
= 294 ÷ 12 : 12 ÷ 12 = 24.5 : 1

Once you've found the surface area of each, set them as a ratio then divide to get the ratio in the form n:1.

<u>24.5 : 1</u>

[4 marks]

Exam Questions

2   The diagram below shows the plan view, and the front and side elevations of a shape made from identical cubes.

How many cubes make up the shape?

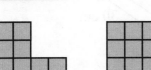 Plan view

Front elevation   Side elevation

....
[2 marks]

3   The diagram below shows the plan, the front elevation and the side elevation of a prism.

Plan view

Draw a sketch of the solid prism on the grid below.

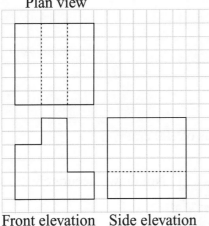

Front elevation   Side elevation

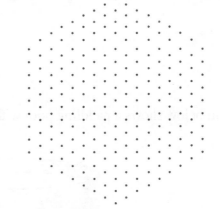

[2 marks]

4   Volume of sphere $= \frac{4}{3}\pi r^3$.
    Find the volume of the sphere on the right.
    Give your answer to 3 significant figures.

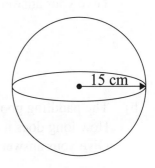

15 cm

.... cm^3
[2 marks]

Exam Questions

5   The tank shown in the diagram below is completely filled with water.

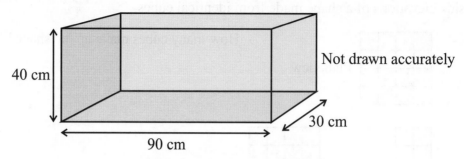

Not drawn accurately

40 cm

90 cm

30 cm

a)   Calculate the volume of water in the tank. **②**

.... cm³
[2 marks]

b)   The water from this tank is then poured into a second tank with length 120 cm. **④**
The depth of the water is 18 cm. What is the width of the second tank?

.... cm
[2 marks]

6   The diagram below shows a paddling pool with a radius of 100 cm. **⑤**

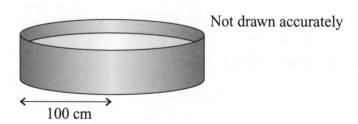

Not drawn accurately

100 cm

a)   What is the volume of water in the paddling pool when it is filled to a depth of 40 cm?
Give your answer in terms of π.

.... cm³
[2 marks]

b)   The paddling pool is filled at a rate of 300 cm³ per second.
How long does it take to fill the pool to a depth of 40 cm?
Give your answer to the nearest minute.

.... minutes
[2 marks]

Revision Questions for Section Five

Now you've finished Section Five, try out your new <u>shape skills</u>...

- Try these questions and <u>tick off each one</u> when you <u>get it right</u>.
- When you're <u>completely happy</u> with a topic, tick that off too.

For even more practice, try the Sudden Fail Quiz for Section Five — just scan this QR code!

Section Five Quiz

2D Shapes (p103-107) ☑

1) For each of the letters shown, write down how many lines of symmetry they have and their order of rotational symmetry. **H Z T N E X S** ☑

2) Write down four properties of an isosceles triangle. ☑

3) How many lines of symmetry does a rhombus have? What is its order of rotational symmetry? ☑

4) What are congruent and similar shapes? ☑

5) Look at the shapes A-G on the right and write down the letters of:
   a) a pair of congruent shapes,
   b) a pair of similar shapes.

   A B C D
   E F G

6) These two triangles are similar.
   Write down the values of *b* and *y*. ☑

   b° 6 cm 89° 3 cm
   8 cm 54° 4 cm 54°
   37° 10 cm 37° *y* cm

Transformations (p108-109) ☑

7) Describe the transformation that maps:
   a) Shape A onto shape B
   b) Shape A onto shape C ☑

8) Carry out the following transformations on the triangle X, which has vertices (1, 1), (4, 1) and (2, 3):
   a) a rotation of 90° clockwise about (1, 1)   b) a translation by the vector $\binom{-3}{-4}$
   c) an enlargement of scale factor 2, centre (1, 1) ☑

Perimeter and Area (p112-113) ☑

9) Write down the formula for finding the area of a trapezium. ☑

10) Find the area of a parallelogram with base 9 cm and vertical height 4 cm. ☑

11) Find the area of the shape on the right. ☑

    3 cm 8 cm 5 cm

12) Find, to 2 decimal places, the area and circumference of a circle with radius 7 cm. ☑

13) Draw a circle and label an arc, a sector, a chord and a segment. ☑

14) Find, to 2 decimal places, the area and perimeter of a quarter circle with radius 3 cm. ☑

3D Shapes — Surface Area, Volume and Projections (p116-121) ☑

15) Write down the number of faces, edges and vertices for the following 3D shapes:
    a) a square-based pyramid   b) a cone   c) a triangular prism ☑

16) Find the surface area of a cube with side length 5 cm. ☑

17) Find, to 1 decimal place, the surface area of a cylinder with height 8 cm and radius 2 cm. ☑

18) Write down the formula for the volume of a cylinder with radius *r* and height *h*. ☑

19) A pentagonal prism has a cross-sectional area of 24 cm² and a length of 15 cm. Find its volume. ☑

20) a) Find the volume of the cylinder on the right (to 2 d.p.).   9 cm
    b) How long will it take to fill the cylinder with water
        if the water is flowing at 1.5 litres per minute?
        Give your answer in seconds to 1 d.p.   2 cm ☑

21) On squared paper, draw the front elevation (from the direction of the arrow),
    side elevation and plan view of the shape on the right. ☑

Angle Basics

*Before we really get going with the thrills and chills of angles and geometry, there are a few things you need to know. Nothing too scary — just some **special angles** and some **fancy notation**.*

Fancy **Angle Names**

Some angles have special names. You might have to <u>identify</u> these angles in the exam.

<u>ACUTE</u> angles

Sharp pointy ones
(less than 90°)

<u>RIGHT</u> angles

Square corners
(exactly 90°)

<u>OBTUSE</u> angles

Flatter ones
(between 90° and 180°)

<u>REFLEX</u> angles

Ones that bend
back on themselves
(more than 180°)

Measuring Angles with a **Protractor**

1) <u>ALWAYS</u> position the protractor with the <u>base line</u> of it along one of the lines as shown here:

2) Count the angle in <u>10° STEPS</u> from the <u>start line</u> right round to the other line over there.

Start line

Check your measurement by looking at it. If it's between a right angle and a straight line, it's between 90° and 180°.

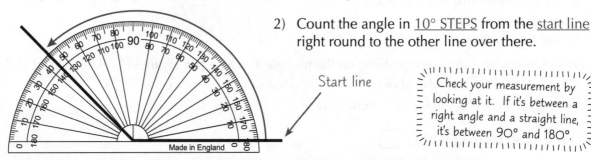

<u>DON'T JUST READ A NUMBER OFF THE SCALE</u> — chances are it'll be the wrong one because there are <u>TWO scales</u> to choose from.

The answer here is 135° (NOT 45°) which you will only get right if you start counting 10°, 20°, 30°, 40° etc. from the <u>start line</u> until you reach the <u>other line</u>.

Three-Letter Angle Notation

The best way to say which angle you're talking about in a diagram is by using <u>THREE letters</u>.

For example in the diagram, angle ACB = 25°.

1) The <u>middle letter</u> is where the angle is.
2) The <u>other two letters</u> tell you which two lines enclose the angle.

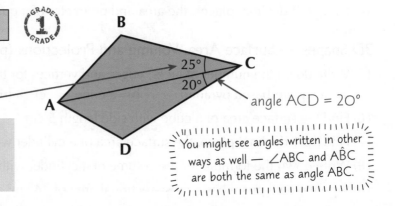

angle ACD = 20°

You might see angles written in other ways as well — ∠ABC and AB̂C are both the same as angle ABC.

Four fancy angle names to learn — acute, right, obtuse and reflex

In the exams, it's pretty likely that angles will be referred to using three-letter notation — so make sure you know how to use it. And learn the four special angles too. Now have a go at this Exam Practice Question.

Q1   a) An angle measures 66°. What type of angle is this?   [1 mark]
         b) Measure ∠ADC on the diagram above.   [1 mark]

Five Angle Rules

If you know all these rules thoroughly, you'll at least have a fighting chance of working out problems with lines and angles. If you don't — you've no chance. Sorry to break it to you like that.

5 Simple Rules — that's all

1) Angles in a <u>triangle</u> add up to <u>180º</u>.

$$a + b + c = 180°$$

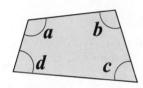

There's a nice proof of this (using <u>parallel lines</u>) on the next page.

2) Angles on a <u>straight line</u> add up to <u>180º</u>.

$$a + b + c = 180°$$

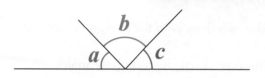

3) Angles in a <u>quadrilateral</u> add up to <u>360º</u>.

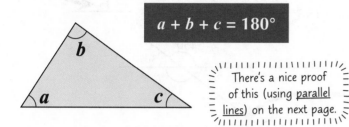

Remember that a quadrilateral is a 4-sided shape.

$$a + b + c + d = 360°$$

You can <u>see why</u> this is if you split the quadrilateral into <u>two triangles</u> along a <u>diagonal</u>. Each triangle has angles adding up to 180°, so the two together have angles adding up to 180° + 180° = 360°.

4) Angles <u>round a point</u> add up to <u>360º</u>.

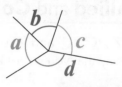

$$a + b + c + d = 360°$$

5) <u>Isosceles triangles</u> have <u>2 sides the same</u> and <u>2 angles the same</u>.

These dashes indicate two sides the same length.

These angles are the same.

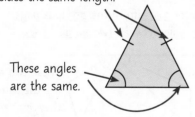

In an isosceles triangle, you only need to know <u>one angle</u> to be able to find the other two.

EXAMPLE **Find the size of angle *x*.**

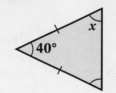

180° − 40° = 140°

<u>The two angles on the right are the same</u> (they're both *x*) and they must add up to 140°, so 2*x* = 140°, which means *x* = 70°.

Five simple rules, make sure you LEARN THEM...

The basic facts are pretty easy really, but examiners like to combine them in questions to confuse you. There are some examples of questions like these on p.129, but have a go at this one as a warm-up.

Q1   Find the size of the angle marked *x*.   [2 marks]

72°

x

Pythagoras' Theorem

Pythagoras' Theorem is dead important — make sure you learn how to use it.

Pythagoras' Theorem is Used on Right-Angled Triangles

Pythagoras' theorem only works for <u>RIGHT-ANGLED TRIANGLES</u>.
It uses <u>two sides</u> to find the <u>third side</u>.

The formula for Pythagoras' theorem is:

$$a^2 + b^2 = c^2$$

short sides long side
(hypotenuse)

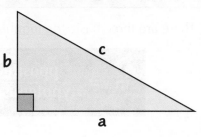

The trouble is, the formula can be quite difficult to use. <u>Instead</u>, it's a lot better to just <u>remember</u> these <u>three simple steps</u>, which work every time:

1) SQUARE THEM — <u>SQUARE THE TWO NUMBERS</u> that you are given (use the x^2 button if you've got your calculator).

2) ADD or SUBTRACT — To find the <u>longest side</u>, <u>ADD</u> the two squared numbers. $a^2 + b^2 = c^2$
To find <u>a shorter side</u>, <u>SUBTRACT</u> the smaller from the larger. $c^2 - b^2 = a^2$

3) SQUARE ROOT — Once you've got your answer, take the <u>SQUARE ROOT</u> (use the $\sqrt{}$ button on your calculator). $c = \sqrt{a^2 + b^2}$
$a = \sqrt{c^2 - b^2}$

EXAMPLES

1. Find the length of side PQ in this triangle.

1) <u>Square</u> them: $a^2 = 12^2 = 144$, $b^2 = 5^2 = 25$

2) You want to find the <u>longest side</u>, so <u>ADD</u>: $a^2 + b^2 = c^2$
$144 + 25 = 169$

3) <u>Square root</u>: $c = \sqrt{169} = 13$ cm

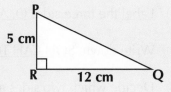

Always check the answer's sensible — 13 cm is longer than the other two sides, but not too much longer, so it seems OK.

2. Find the length of SU to 1 decimal place.

1) <u>Square</u> them: $b^2 = 3^2 = 9$, $c^2 = 6^2 = 36$

2) You want to find <u>a shorter side</u>, so <u>SUBTRACT</u>: $c^2 - b^2 = a^2$
$36 - 9 = 27$

3) <u>Square root</u>: $a = \sqrt{27} = 5.196...$
$= 5.2$ m (to 1 d.p.)

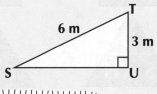

Check the answer is <u>sensible</u> — yes, it's a bit shorter than the longest side.

Use Pythagoras' theorem to find lengths in right-angled triangles

This is one of the most famous of all maths theorems and it'll probably be in your exam at some point. You really need to learn the method and practise plenty of questions. So here are two to get you started.

Q1 Look at triangle ABC on the right.
Find the length of side AC correct to 1 decimal place. [3 marks]

Q2 A 4 m long ladder leans against a wall. Its base is 1.2 m from the wall. How far up the wall does the ladder reach?
Give your answer to 1 decimal place. [3 marks]

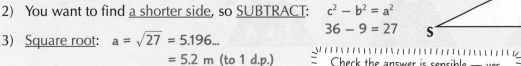

Q2 Video Solution

Section Six — Angles and Geometry

Trigonometry — Sin, Cos, Tan

Trigonometry — *it's clever stuff. The three trig formulas are used on right-angled triangles to:*
a) find an unknown side if you know a side and an angle, or b) find an angle if you know two lengths.

The **3 Trigonometry Formulas**

There are three basic <u>trig formulas</u> — each one links <u>two sides and an angle</u> of a <u>right-angled triangle</u>.

$$\text{Sin } x = \frac{\text{Opposite}}{\text{Hypotenuse}}$$

$$\text{Cos } x = \frac{\text{Adjacent}}{\text{Hypotenuse}}$$

$$\text{Tan } x = \frac{\text{Opposite}}{\text{Adjacent}}$$

- The <u>Hypotenuse</u> is the <u>LONGEST SIDE</u>.
- The <u>Opposite</u> is the side <u>OPPOSITE</u> the angle <u>being used</u> (*x*).
- The <u>Adjacent</u> is the (other) side <u>NEXT TO</u> the angle <u>being used</u>.

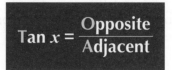

Opposite (O) Hypotenuse (H) Adjacent (A) *x*

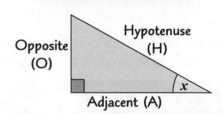

Formula Triangles Make Things Easier

There's more about formula triangles on p.98 if you need to jog your memory.

A great way to tackle trig questions is to convert the formulas into <u>formula triangles</u>.
Then you can use the <u>same method every time</u>, no matter which side or angle is being asked for.

1) <u>Label</u> the three sides <u>O, A and H</u> (Opposite, Adjacent and Hypotenuse).

2) Write down '<u>SOH CAH TOA</u>'.

3) Decide which <u>two sides</u> are <u>involved</u>: O,H A,H or O,A
and choose <u>SOH</u>, <u>CAH</u> or <u>TOA</u> accordingly.

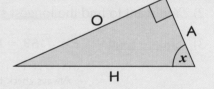

O A *x* H

4) Turn the one you choose into a <u>FORMULA TRIANGLE</u>:

S O H
$$\frac{O}{S \times H}$$

C A H
$$\frac{A}{C \times H}$$

T O A
$$\frac{O}{T \times A}$$

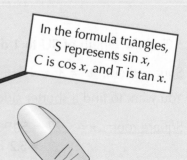

In the formula triangles, S represents sin *x*, C is cos *x*, and T is tan *x*.

5) <u>Cover up</u> the thing you want to find with your finger,
and write down whatever is left showing.

6) <u>Stick in the numbers</u> and work it out using the
[sin], [cos] and [tan] buttons on your <u>calculator</u>.

If you're finding an <u>angle</u>, you'll need to add an extra step to find the <u>inverse</u>. See next page.

H = longest, O = opposite, A = next to, and remember SOH CAH TOA
You really need to know all of this stuff off by heart — so go over this page a few more times.
Make sure you've got those formulas firmly lodged in your mind and all ready to reel off in the exam.

Trigonometry — Examples

Here are some lovely examples using the method from the previous page to help you through the trials of trig.

Examples:

---

1 **Find the length of x in the triangle to the right.**

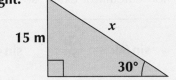

1) <u>Label</u> the sides ———

2) <u>Write down</u> ——— (SOH) CAH TOA

3) <u>O</u> and <u>H</u> involved

4) Write down the <u>formula triangle</u> ———

5) <u>You want H</u> so <u>cover it up</u> to give ——— $H = \dfrac{O}{S}$

6) <u>Put in</u> the <u>numbers</u>

    [15] [÷] [sin] [30] [=]

    $x = \dfrac{15}{\sin 30°} = \dfrac{15}{0.5} = 30\ \text{m}$

---

2 **Find the angle x in the triangle to the right.**

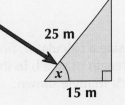

1) <u>Label</u> the sides ———

2) <u>Write down</u> ——— SOH (CAH) TOA

3) <u>A</u> and <u>H</u> involved

4) Write down the <u>formula triangle</u>

5) You want the <u>angle</u> so <u>cover up C</u> to give ——— $C = \dfrac{A}{H}$

6) Put in the numbers

    [15] [÷] [25] [=]

    $\cos x = \dfrac{15}{25} = 0.6$

7) Find the <u>inverse</u>.

    [shift] [cos] [0.6] [=]

    $\Rightarrow x = \cos^{-1}(0.6) = 53.1301...°$

    $= 53.1°$ (1 d.p.)

> When you're finding an <u>angle</u> you'll have to find the <u>INVERSE</u> at the end. Press [SHIFT] or [2ndF], followed by sin, cos or tan — your calculator will display $\sin^{-1}$, $\cos^{-1}$ or $\tan^{-1}$.

You need to have learnt all seven steps

In example 2 you can see the six steps from the previous page, plus an extra one put into action. Remember — you only need to do step 7 if you're finding an angle. Now use the seven steps to answer these questions.

Q1 Find the length of y and give your answer to 2 decimal places.

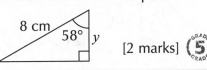

[2 marks]

Q2 Find the value of x and give your answer to 1 decimal place.

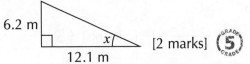

[2 marks]

Q2 Video Solution

Trigonometry — Common Values

Now that you're in the swing of trigonometry questions it's time to put those calculators away. Sorry.

Learn these **Common Trig Values**

The tables below contain a load of <u>useful trig values</u>. You might get asked to work out some <u>exact</u> trig answers in your non-calculator exam, so having these in your brain will come in handy.

$$\sin 30° = \frac{1}{2} \qquad \sin 60° = \frac{\sqrt{3}}{2} \qquad \sin 45° = \frac{1}{\sqrt{2}}$$

$$\cos 30° = \frac{\sqrt{3}}{2} \qquad \cos 60° = \frac{1}{2} \qquad \cos 45° = \frac{1}{\sqrt{2}}$$

$$\tan 30° = \frac{1}{\sqrt{3}} \qquad \tan 60° = \sqrt{3} \qquad \tan 45° = 1$$

If you're asked for <u>exact</u> answers, <u>don't</u> convert them to decimals at the end.

$$\sin 0° = 0 \qquad \cos 0° = 1 \qquad \tan 0° = 0$$

$$\sin 90° = 1 \qquad \cos 90° = 0$$

Have a look at the <u>examples</u> below —
they might help cement a few values into your head.

EXAMPLES

1. Without using a calculator, find the exact length of side b in the right-angled triangle shown.

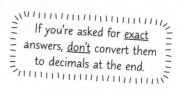

1) It's a right-angled triangle so use SOH CAH TOA to pick the correct <u>trig formula</u> to use.

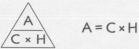

$$A = C \times H$$

2) Put in the <u>numbers</u> from the diagram in the question.

$$b = \cos 30° \times 7$$

3) You know the <u>value</u> of <u>cos 30°</u>, so <u>substitute</u> this in.

$$b = \frac{\sqrt{3}}{2} \times 7 = \frac{7\sqrt{3}}{2} \text{ cm}$$

2. Without using a calculator, show that $\cos 60° + \sin 30° = 1$

Put in the right values for $\cos 60°$ and $\sin 30°$, then do the sum.

$$\cos 60° = \frac{1}{2} \qquad \sin 30° = \frac{1}{2}$$

$$\cos 60° + \sin 30° = \frac{1}{2} + \frac{1}{2} = 1$$

Learn these common trig values off by heart

There are a lot of angles to learn on this page but you need to know them all. If a trigonometry question asks for an <u>exact</u> answer, you'll need to use them. When you're ready, try this Exam Practice Question.

Q1   Find the exact length of side *x*.

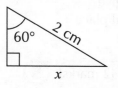

[2 marks]

Q2   Find the exact length of side *y*.

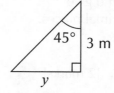

[2 marks]

Q2 Video Solution

Vectors — Theory

*Vectors represent a movement of a certain **size** in a certain **direction**.*
They might seem a bit weird at first, but there are really just a few facts to get to grips with...

The Vector **Notations**

There are several ways to <u>write</u> vectors...

They're shown on
a diagram by an <u>arrow</u>.

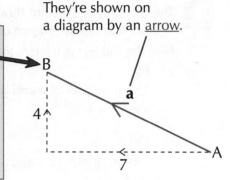

1) <u>Column</u> vectors: $\begin{pmatrix} 2 \\ -5 \end{pmatrix}$—— 2 units right —— 5 units down $\begin{pmatrix} -7 \\ 4 \end{pmatrix}$—— 7 units left —— 4 units up

2) **a** —— <u>exam questions</u> use <u>bold</u> like this

3) <u>a</u> —— <u>you</u> should always <u>underline</u> them

4) $\overrightarrow{AB}$ —— this means the vector <u>from point A to point B</u>

Multiplying a Vector **by a Number**

1) Multiplying a vector by a <u>positive</u> number <u>changes</u> the vector's <u>size</u> but <u>not its direction</u>.

2) If the number's <u>negative</u> then the <u>direction gets switched</u>.

Vectors that are <u>multiples</u> of each other are <u>parallel</u>.

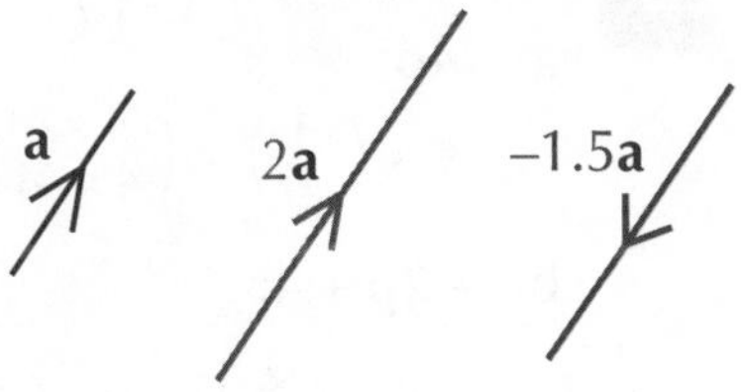

a   2**a**   −1.5**a**

Adding and **Subtracting** Vectors

You can describe movements between points by <u>adding and subtracting known vectors</u>.

"<u>a</u> + <u>b</u>" means 'go along <u>a</u> then <u>b</u>'.

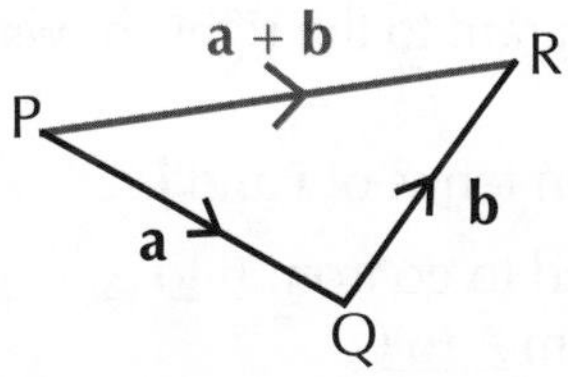

In the diagrams,
$\overrightarrow{PR}$ = <u>a</u> + <u>b</u> and
$\overrightarrow{XZ}$ = <u>c</u> − <u>d</u>.

"<u>c</u> − <u>d</u>" means 'go along <u>c</u> then backwards along <u>d</u>' (the <u>minus</u> sign means go the <u>opposite</u> way).

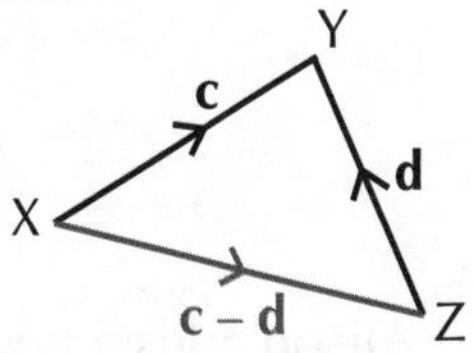

When adding <u>column vectors</u>, add the top to the top and the bottom to the bottom. The same goes when subtracting.

$$\begin{pmatrix} 3 \\ -1 \end{pmatrix} + \begin{pmatrix} 5 \\ 3 \end{pmatrix} = \begin{pmatrix} 3+5 \\ -1+3 \end{pmatrix} = \begin{pmatrix} 8 \\ 2 \end{pmatrix}$$

That's three vital vector facts done

But they're only really 'done' if you've learnt them. So be sure you know how vectors can be written, can multiply a vector by a number and can add and subtract vectors. Now have a go at these lovely questions.

Q1   Calculate $\begin{pmatrix} 4 \\ 7 \end{pmatrix} - \begin{pmatrix} 3 \\ 2 \end{pmatrix}$.   [1 mark]

Q2   Find $\overrightarrow{LN}$ on the diagram to the right.   [1 mark]

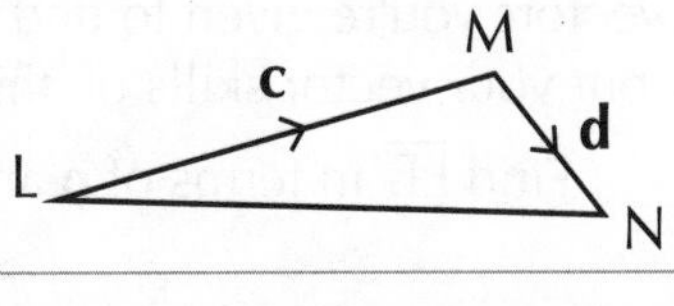

Vectors — Examples

Here's a full page of worked examples to get you in the mood for some tasty vector questions of your own.

Examples

> When you're going in the opposite direction to the vector, <u>reverse the sign</u>.

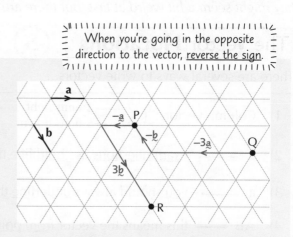

1 The diagram to the right shows an isometric grid. Vectors **a** and **b** are shown on the grid.

Find the following vectors in terms of **a** and **b**:

a) $\overrightarrow{PR}$

Use multiples of $\underset{\sim}{a}$ and $\underset{\sim}{b}$ to get from <u>P to R</u>.

$\overrightarrow{PR} = -\underset{\sim}{a} + 3\underset{\sim}{b}$

b) $\overrightarrow{QP}$

Use multiples of $\underset{\sim}{a}$ and $\underset{\sim}{b}$ to get from <u>Q to P</u>.

$\overrightarrow{QP} = -3\underset{\sim}{a} - \underset{\sim}{b}$

---

2 Given that $\mathbf{p} = \begin{pmatrix} 4 \\ -1 \end{pmatrix}$ and $\mathbf{r} = \begin{pmatrix} -3 \\ 2 \end{pmatrix}$ find:

a) $4\mathbf{p}$

$$4 \times \begin{pmatrix} 4 \\ -1 \end{pmatrix} = \begin{pmatrix} 4 \times 4 \\ 4 \times -1 \end{pmatrix} = \begin{pmatrix} 16 \\ -4 \end{pmatrix}$$

b) $-2\mathbf{p} + 4\mathbf{r}$

$$-2 \times \begin{pmatrix} 4 \\ -1 \end{pmatrix} + 4 \times \begin{pmatrix} -3 \\ 2 \end{pmatrix} = \begin{pmatrix} -2 \times 4 \\ -2 \times -1 \end{pmatrix} + \begin{pmatrix} 4 \times -3 \\ 4 \times 2 \end{pmatrix}$$

$$= \begin{pmatrix} -8 \\ 2 \end{pmatrix} + \begin{pmatrix} -12 \\ 8 \end{pmatrix} = \begin{pmatrix} -20 \\ 10 \end{pmatrix}$$

---

3 The diagram to the right shows the vectors $\overrightarrow{BA}$ and $\overrightarrow{AC}$.

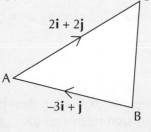

a) Find $\overrightarrow{BC}$ in terms of **i** and **j**.

You need to go from <u>B to A</u>, then from <u>A to C</u>.

$\overrightarrow{BA} = -3\underset{\sim}{i} + \underset{\sim}{j}$

$\overrightarrow{AC} = 2\underset{\sim}{i} + 2\underset{\sim}{j}$

$\overrightarrow{BC} = -3\underset{\sim}{i} + \underset{\sim}{j} + 2\underset{\sim}{i} + 2\underset{\sim}{j}$

$= -\underset{\sim}{i} + 3\underset{\sim}{j}$

b) Alison adds a new line, BD, to the diagram. Given that $\overrightarrow{BD}$ is in the opposite direction to $\overrightarrow{BC}$ and is twice the length of $\overrightarrow{BC}$, find $\overrightarrow{BD}$ in terms of **i** and **j**.

$\overrightarrow{BD}$ is in the <u>opposite</u> direction to $\overrightarrow{BC}$, so <u>reverse</u> the sign. $\longrightarrow \underset{\sim}{i} - 3\underset{\sim}{j}$

$\overrightarrow{BD}$ is also <u>twice</u> the length of $\overrightarrow{BC}$, so <u>multiply by 2</u>. $\longrightarrow 2\underset{\sim}{i} - 6\underset{\sim}{j}$

So $\overrightarrow{BD} = 2\underset{\sim}{i} - 6\underset{\sim}{j}$

You could get similar questions to these on the exam

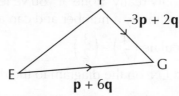

So learn how to do these types of questions — just use the vectors you're given to find what you're asked for. Test out your vector skills on this practice question.

Q1   Find $\overrightarrow{EF}$ in terms of **p** and **q**.   [2 marks]

Q1 Video Solution

Warm-up and Worked Exam Questions

There are a lot of different ideas to take in from this mini-section, so here's a bit of a warm-up to get you into the swing of things before the exam questions on the next page.

Warm-up Questions

1) Find the length of the side AB. Give your answer to 1 decimal place.

2) Find the length of z correct to 2 decimal places.

3) Without using a calculator, find the exact length of side x in this triangle.

4) Without a calculator, find the value of tan 45° – cos 60°.

5) Given that $\mathbf{b} = \begin{pmatrix} -5 \\ 3 \end{pmatrix}$ and $\mathbf{c} = \begin{pmatrix} 1 \\ 6 \end{pmatrix}$ find:

    a) $3\mathbf{b}$ b) $-2\mathbf{c}$ c) $3\mathbf{c} - \mathbf{b}$

6) For the diagram on the right, find $\overrightarrow{EF}$ in terms of $\mathbf{a}$ and $\mathbf{b}$.

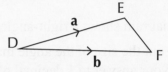

Worked Exam Questions

Some more exam questions with the answers written in — just what the doctor ordered.

1   A rectangle has a height of 3 cm and a diagonal length of 5 cm. **(4)**

Calculate the area of the rectangle.

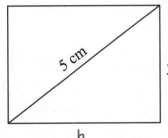

Not to scale

$5^2 = 3^2 + b^2$ ———— Use Pythagoras'
$b^2 = 25 - 9 = 16$ Theorem
$b = \sqrt{16}$
$b = 4$ cm
Area of rectangle = 3 cm × 4 cm
                = 12 cm²

.....12..... cm²
[4 marks]

2   The diagram shows a right-angled triangle. **(5)**

Work out the exact length of the side marked y.

You want to find the opposite length, so cover up O.

$y = \tan 60° × 4$

$y = \sqrt{3} × 4 = 4\sqrt{3}$ m

60°   Not drawn accurately

4 m

y

.....$4\sqrt{3}$..... m
[2 marks]

Exam Questions

3   An isosceles triangle has a base of 10 cm. Its other two sides are both 13 cm long. **(4)**
    Calculate the height of the triangle.

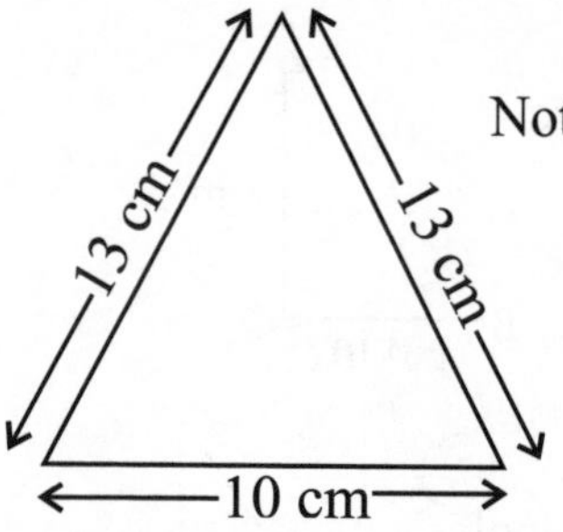

Not to scale

Think about whether you're trying to find the hypotenuse or one of the shorter sides before using Pythagoras' Theorem.

13 cm   13 cm

←——10 cm——→

.... cm
[3 marks]

4   The diagram shows a right-angled triangle. **(5)**

    Work out the size of the angle marked x.
    Give your answer to 1 decimal place.

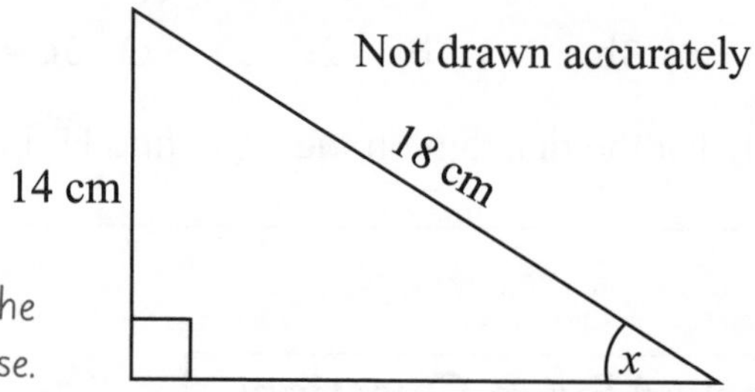

Not drawn accurately

14 cm   18 cm

The sides involved here are the Opposite and the Hypotenuse.

x

...........................°
[2 marks]

5   ABC is a triangle. $\overrightarrow{AB} = 2\mathbf{c}$ and $\overrightarrow{BC} = 2\mathbf{d}$. L is the midpoint of AC. **(5)**

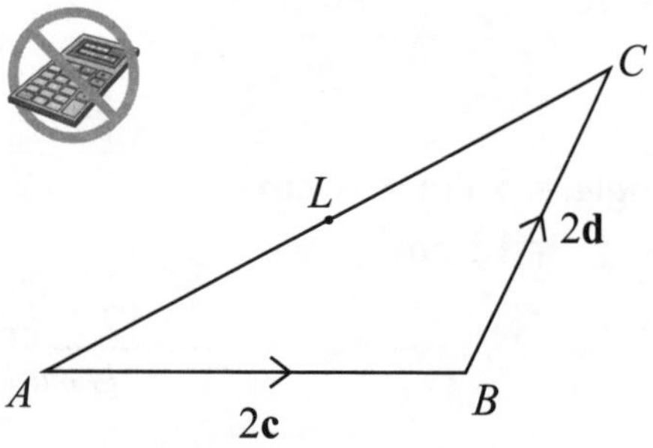

Write in terms of $\mathbf{c}$ and $\mathbf{d}$:

C

L

2**d**

a) $\overrightarrow{AC}$

....
[2 marks]

A   B
   2**c**

b) $\overrightarrow{AL}$

The midpoint is found halfway along the line.

Not drawn accurately

....
[2 marks]

c) $\overrightarrow{BL}$

....
[2 marks]

Revision Questions for Section Six

Look — a chance to use your artistic skills while doing maths...
- Try these questions and <u>tick off each one</u> when you <u>get it right</u>.
- When you're <u>completely happy</u> with a topic, tick that off too.

For even more practice, try the Sudden Fail Quiz for Section Six — just scan this QR code!

Section Six Quiz

Angles and Geometry Problems (p126-129) ☑

1) What is the name for an angle larger than 90° but smaller than 180°? ☑

2) What do angles in a quadrilateral add up to? ☑

3) Find the missing angles in the diagrams below.

   a)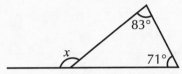
   b)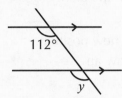
   c)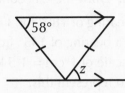

4) Given that angle DAC = 70°, work out angle CED. ☑

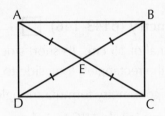

Angles in Polygons (p130) ☑

5) Find the exterior angle of a regular hexagon. ☑

6) Find the sum of interior angles in a regular octagon. ☑

7) Find the interior angle of a regular 20-sided polygon. ☑

Constructions and Loci (p133-136) ☑

8) Construct triangle XYZ, where XY = 5.6 cm, XZ = 7.2 cm and angle YXZ = 55°. ☑

9) What shape does the locus of points that are a fixed distance from a given point make? ☑

10) Draw a horizontal line with a length of 8 cm.
    Draw the locus of points exactly 2 cm away from the line. ☑

11) Construct an accurate 90° angle. ☑

12) Draw a square with sides of length 6 cm and label it ABCD. Shade the region
    that is nearer to AB than CD and less than 4 cm from vertex A. ☑

Bearings (p137) ☑

13) Using the diagram on the right, find the bearing of Y from X. ☑

14) Look at the diagram to the right.
    Tom wants to travel from point A to point B.
    Find the bearing he should travel on. ☑

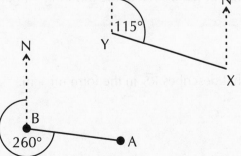

Revision Questions for Section Six

Maps and Scale Drawings (p138-139)

15) The scale on a map is 1 cm = 4 km. On the map, Leadz is 6.5 cm away from Horrowgate.
How far is this in real life?

16) The garden plan on the right has a scale of 1:200.
Allen wants to build a square shed in the top right corner
of the garden and a rectangular pond in the bottom right.
The shed will measure 4 m × 4 m and the pond will
measure 6 m × 2 m. Draw these accurately on the plan.

17) John travels on a bearing of 180° for 3 km.
He then travels on a bearing of 145° from his new position
for 6 km. Using a scale of 1 cm = 1.5 km, draw an
accurate diagram to represent this.

Pythagoras and Trigonometry (p143-146)

18) A rectangle has a diagonal of 15 cm. Its short side is 4 cm.
Calculate the length of the rectangle's long side to 1 d.p.

19) Write down the three trigonometry formula triangles.

20) Find the size of angle x in triangle ABC to 1 d.p.

21) Work out the value of x in triangle PQR to 1 d.p.

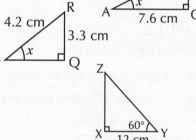

22) Find the exact length of side XZ in triangle XYZ
without using your calculator.

23) Without using your calculator, show that $\tan 45° + \sin 60° = \dfrac{2+\sqrt{3}}{2}$

Vectors (p147-148)

24) $\underset{\sim}{a}$ and $\underset{\sim}{b}$ are column vectors, where $\underset{\sim}{a} = \begin{pmatrix} 4 \\ -2 \end{pmatrix}$ and $\underset{\sim}{b} = \begin{pmatrix} 7 \\ 6 \end{pmatrix}$.
   a) Find $\underset{\sim}{a} - \underset{\sim}{b}$
   b) Find $5\underset{\sim}{a}$

25) The diagram to the right shows a grid of unit squares.
Find the vector $\underset{\sim}{c}$ in terms of $\underset{\sim}{a}$ and $\underset{\sim}{b}$.

26) Find the vector that describes $\overrightarrow{XY}$ in the form $m\underset{\sim}{a} + n\underset{\sim}{b}$.

27) Find the vector that describes $\overrightarrow{RS}$ in the form $m\underset{\sim}{i} + n\underset{\sim}{j}$.

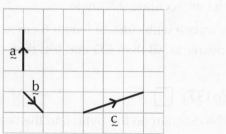

Probability Basics

*A lot of people think **probability** is tough. But learn the **basics** well, and it'll all make sense.*

All **Probabilities** are **Between 0 and 1**

1) Probabilities are <u>always</u> between 0 and 1.
2) The <u>higher</u> the probability of something, the <u>more likely</u> it is.
3) A probability of <u>ZERO</u> means it will <u>NEVER HAPPEN</u>.
4) A probability of <u>ONE</u> means it <u>DEFINITELY WILL HAPPEN</u>.

You can show the probability of something happening on a <u>scale</u> from 0 to 1.
Probabilities can be given as <u>fractions</u>, <u>decimals</u> or <u>percentages</u>.

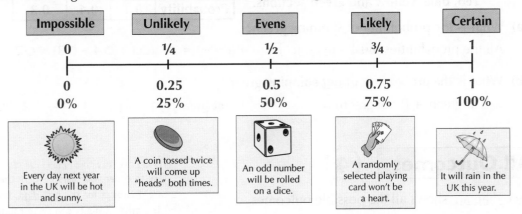

Impossible	Unlikely	Evens	Likely	Certain
0	¼	½	¾	1
0	0.25	0.5	0.75	1
0%	25%	50%	75%	100%
Every day next year in the UK will be hot and sunny.	A coin tossed twice will come up "heads" both times.	An odd number will be rolled on a dice.	A randomly selected playing card won't be a heart.	It will rain in the UK this year.

Use This **Formula** When **All Outcomes** are **Equally Likely**

Use this formula to find probabilities for a <u>fair</u> spinner, coin or dice.
A spinner/coin/dice is 'fair' when it's <u>equally likely</u> to land on <u>any</u> of its sides.

$$\text{Probability} = \frac{\text{Number of ways for something to happen}}{\text{Total number of possible outcomes}}$$

<u>Outcomes</u> are just 'things that could happen'.

EXAMPLE **The picture on the right shows a fair, 8-sided spinner.**

a) **Work out the probability of this spinner landing on green.**

There are <u>8 sides</u> so there are <u>8 possible outcomes</u>.
There are <u>3 ways</u> for it to land on <u>green</u>.

P(green) means 'The probability of the spinner landing on green'.

$$\text{P(green)} = \frac{\text{number of ways for 'green' to happen}}{\text{total number of possible outcomes}} = \frac{3}{8} \text{ (or 0.375)}$$

b) **Which of its four colours is the spinner <u>least likely</u> to land on?**

It's least likely to land on the colour that 'can happen in the <u>fewest ways</u>' — this is the one on the <u>fewest sections</u>.   White

---

A probability of 1 means it's certain to happen

A probability of 0 means it definitely won't happen — the higher the probability, the more likely it is.

Q1 Calculate the probability of the fair spinner on the right:
   a) landing on a 4.   [2 marks]   b) landing on an even number.   [2 marks]
Q2 Show the probabilities in Q1 on a scale from 0 to 1.   [1 mark]

More Probability

Probabilities **Add Up To 1**

1) If <u>only one</u> possible result can happen at a time, then the probabilities of <u>all</u> the results <u>add up to 1</u>.

Probabilities always ADD UP to 1

2) So since something must either <u>happen</u> or <u>not happen</u> (i.e. <u>only one</u> of these can happen at a time):

P(event happens) + P(event doesn't happen) = 1

EXAMPLE A spinner has different numbers of red, blue, yellow and green sections.

Colour	red	blue	yellow	green
Probability	0.1	0.4	0.3	

a) **What is the probability of spinning green?**

All the probabilities must <u>add up to 1</u>.   P(green) = 1 − (0.1 + 0.4 + 0.3) = 0.2

b) **What is the probability of <u>not</u> spinning green?**

P(green) + P(not green) = 1   P(not green) = 1 − P(green) = 1 − 0.2 = 0.8

Listing **All Outcomes**

A <u>sample space diagram</u> shows all the possible outcomes.

Try to order your lists — here there are 3 choices for the first digit, then the other 2 digits can swap round.

1) It can just be a <u>simple list</u>...
   E.g. Find all the 3-digit numbers that include the digits 1, 2 and 3.   1<u>2</u>3, 1<u>3</u>2, 2<u>1</u>3, 2<u>3</u>1, 3<u>2</u>1, 3<u>1</u>2

2) Or you can draw a <u>two-way table</u> if there are <u>two activities</u> going on
   (e.g. two coins being tossed, or a dice being thrown and a spinner being spun).

EXAMPLE   **The spinner on the right is spun twice, and the scores added together.**

a) **Complete this sample space diagram showing all the possible outcomes.**

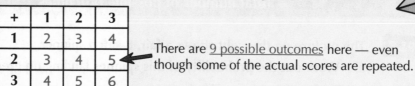

+	1	2	3
1	2	3	4
2	3	4	5
3	4	5	6

There are <u>9 possible outcomes</u> here — even though some of the actual scores are repeated.

b) **Find the probability of spinning a total of 3.**

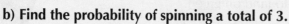

$$P(\text{total of 3}) = \frac{\text{ways to score 3}}{\text{total number of possible outcomes}} = \frac{2}{9}$$

There are 2 ways to score 3.

c) **Find the probability of spinning a total of 4 or more.**

$$P(\text{total of 4 or more}) = \frac{\text{ways to score 4 or more}}{\text{total number of possible outcomes}} = \frac{6}{9} = \frac{2}{3}$$

There are 6 ways to score either 4, 5 or 6.

List all possible outcomes in a methodical way

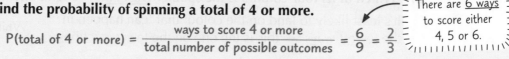

Be careful when you're listing outcomes — you don't want to lose easy marks by missing one out.

Q1 Two fair dice are thrown, and the scores added together. By drawing a sample space diagram,
   a) find the probability of throwing a total of 7,   [3 marks]
   b) find the probability of throwing any total except 7.   [1 mark]

Q1 Video Solution

Probability Experiments

*The formula on page 153 only works when the outcomes are equally likely. If they're **not equally likely**, you can use the results from experiments to **estimate** the probability of each outcome.*

Do the Experiment **Again** and **Again**...

You need to do an experiment <u>over and over again</u> and count how many times each outcome happens (its <u>frequency</u>). Then you can calculate the <u>relative frequency</u> using this formula:

$$\text{Relative frequency} = \frac{\text{Frequency}}{\text{Number of times you tried the experiment}}$$

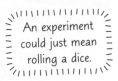
An experiment could just mean rolling a dice.

You can use the <u>relative frequency</u> of a result as an <u>estimate</u> of its <u>probability</u>.

EXAMPLE

The spinner on the right was spun 100 times. Use the results in the table below to estimate the probability of getting each of the scores.

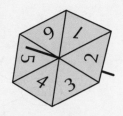

Score	1	2	3	4	5	6
Frequency	3	14	41	20	18	4

<u>Divide</u> each of the frequencies by 100 to find the <u>relative frequencies</u>.

Score	1	2	3	4	5	6
Relative Frequency	$\frac{3}{100} = 0.03$	$\frac{14}{100} = 0.14$	$\frac{41}{100} = 0.41$	$\frac{20}{100} = 0.2$	$\frac{18}{100} = 0.18$	$\frac{4}{100} = 0.04$

The <u>MORE TIMES</u> you do the experiment, the <u>MORE ACCURATE</u> your estimate of the probability should be. E.g. if you spun the above spinner <u>1000 times</u>, you'd get a <u>better</u> estimate of the probability for each score.

Fair or **Biased**?

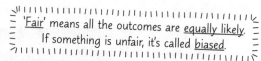
'Fair' means all the outcomes are <u>equally likely</u>. If something is unfair, it's called <u>biased</u>.

1) If the dice/spinner/coin/etc. is <u>fair</u>, then the relative frequencies of the results should <u>roughly match</u> the probabilities you'd get using the formula on p.153.

2) If the relative frequencies are <u>far away</u> from those probabilities, you can say it's probably <u>biased</u>.

EXAMPLE

Do the above results suggest that the spinner is biased?

Yes, because the relative frequency of 3 is much higher than you'd expect, while the relative frequencies of 1 and 6 are much lower.

For a <u>fair</u> 6-sided spinner, you'd expect all the relative frequencies to be about 1 ÷ 6 = 0.17(ish).

More experiments mean a more accurate probability estimate

Remember that relative frequency can only be used to <u>estimate</u> the probability of a result. You can increase the <u>accuracy</u> of your estimate by increasing the number of times you do the experiment.

Q1 This table shows how many times Sekai and Sandro got a free biscuit on their visits to a coffee shop.

	Sekai	Sandro
Visits to coffee shop	20	150
Got a free biscuit	13	117

   a) Based on Sekai's results, estimate the probability of getting a free biscuit at the shop. [2 marks]
   b) Whose results will give a better estimate of the probability? Explain your answer. [1 mark]

Probability Experiments

*OK, I'll admit it, probability experiments aren't as fun as science experiments but they are **useful**.*

Record Results in **Frequency Trees**

When an experiment has two or more steps, you can record the results using a <u>frequency tree</u>.

EXAMPLE

120 GCSE maths students were asked if they would go on to do A-level maths.
- **45 of them said they would go on to do A-level maths.**
- **30 of the students who said they would do A-level maths actually did.**
- **9 of the students who said they wouldn't do A-level maths actually did.**

a) **Complete the frequency tree below.**

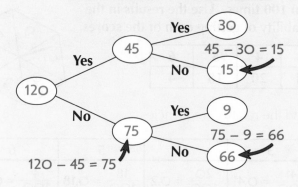

Will they take A-level maths?   Did they take A-level maths?

120
Yes — 45
No — 75 (120 − 45 = 75)
45: Yes — 30
45: No — 15 (45 − 30 = 15)
75: Yes — 9
75: No — 66 (75 − 9 = 66)

b) **A student who said they wouldn't do A-level maths is chosen at random. What is the probability that they <u>did</u> do A-level maths?**

<u>9</u> out of the <u>75 students</u> who said they wouldn't do A-level maths actually did.

So the probability is $\frac{9}{75}$ = 0.12

Use Probability to Find an **"Expected Frequency"**

1) You can <u>estimate</u> how many times you'd <u>expect</u> something to happen if you do an experiment <u>*n* times</u>.
2) This <u>expected frequency</u> is based on the <u>probability</u> of the result happening.

Expected frequency of a result = probability × number of trials

A <u>trial</u> is a single experiment.

EXAMPLE

A person spins the fair spinner on the right 200 times. Estimate how many times it will land on 5.

1) First calculate the probability of the spinner <u>landing on 5</u>.

$$P(\text{lands on 5}) = \frac{\text{ways to land on 5}}{\text{number of possible outcomes}} = \frac{1}{8}$$

2) Then <u>multiply</u> by <u>200</u> spins.

Expected number of 5's = P(lands on 5) × number of trials

$$= \frac{1}{8} \times 200 = 25 \text{ times}$$

If you don't know the probability of a result, fear not...
... you can estimate the probability using the <u>relative frequency</u> of the result in <u>past</u> experiments.

Expected frequency is how many times you'd expect something to happen

Make sure you can remember the formula for expected frequency in the box above. Try to get your head around where you can use frequency trees — they're really useful for experiments with two or more steps.

Q1 Using the frequency tree above, find the probability that a randomly chosen student said they were going to do A-level maths but didn't actually do it. [2 marks]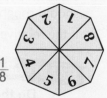

Q2 The spinner above is spun 300 times.
Estimate how many times it will land on an even number. [2 marks]

Warm-up and Worked Exam Questions

These probability basics aren't that difficult, but it's easy to throw away marks by being a little slap-dash with your calculations. So it's important to get loads of practice. Try these questions.

Warm-up Questions

1) Calculate the probability of the fair spinner on the right landing on a red triangle.

2) The probability of rolling a double from two dice rolls is $\frac{1}{6}$.
   What is the probability of not rolling a double?

3) Two fair coins are tossed: a) List all the possible outcomes.
                              b) Find the probability of getting exactly 1 head.

4) Elsa rolled a dice 1000 times and got the results shown in the table below.

Score	1	2	3	4	5	6
Frequency	140	137	138	259	161	165

   Find the relative frequencies for each of the scores 1-6.

5) Using the frequency tree on page 156, find the probability that a randomly chosen student said they were going to do A-level maths and actually did.

6) The spinner on page 156 is spun 500 times.
   Estimate how many times it will land on a factor of 6.

Worked Exam Questions

Take a look at these worked exam questions. They should give you a good idea of how to answer similar questions. You'll usually get at least one probability question in the exam.

1   Sophie records the positions of all the members of her
    football team. The table on the right shows her results.

    A member of the team is chosen at random.
    What is the probability they're a midfielder?
    Give your answer as a decimal.

Position	Frequency
Attacker	6
Midfielder	9
Defender	4
Goalkeeper	1

   There are 6 + 9 + 4 + 1 = 20 people on the team

   P(midfielder) = $\frac{9}{20}$ = 0.45 ⟵ 9 out of the 20 people on
                                          the team are midfielders.

   ......0.45......
   [2 marks]

2   George has a biased 5-sided spinner numbered 1-5.
    The table below shows the probabilities of the spinner landing on numbers 1-4.

Number	1	2	3	4	5
Probability	0.3	0.15	0.2	0.25	

   He spins the spinner 100 times. Estimate the number of times it will land on 5.

   P(landing on a 5) = 1 − (0.3 + 0.15 + 0.2 + 0.25) = 0.1

   In 100 spins, you'd estimate: 0.1 × 100 = 10 fives
                    | \
   Expected frequency = probability × no. of trials

   ......10......
   [2 marks]

Exam Questions

3   There are 10 counters in a bag. 4 of the counters are blue and the rest are red. (2)
    One counter is picked out at random. On the scale below,
    mark with an arrow (↓) the probability that a red counter is picked.

```
      |  |  |  |  |  |  |  |  |  |  |  |  |  |  |  |  |  |  |  |  |
      0                          0.5                          1
```
                                                                    [2 marks]

4   Katie decides to attend two new after-school activities. She can do one on Monday (3)
    and one on Thursday. Below are lists of the activities she could do on these days.

Monday	**Thursday**
Hockey	Netball
Orchestra	Choir
Drama	Orienteering

a) List all nine possible combinations of two activities Katie could try in one week.

                                                                    [2 marks]

Katie randomly picks an activity to do on each day.
Use your answer to part a) to find:

b) the probability that she does hockey on Monday and netball on Thursday,

    ....
                                                                    [1 mark]

c) the probability that she does drama on Monday.

    ....
                                                                    [1 mark]

5   Mani has 3 pieces of homework, English (E), History (H) and Maths (M). (3)
    He has to do all 3 pieces tonight but he can do them in any order.

a) List all the different orders in which he could do his pieces of homework.

    ....
    ....
                                                                    [2 marks]

Mani randomly chooses the order in which to do his pieces of homework.

b) Use your answer to part a) to find the probability that he does his Maths homework
    before his English homework. Give your answer as a fraction in its simplest form.

    ....
                                                                    [1 mark]

Exam Questions

6 Danielle flipped a coin 100 times and predicted the outcome before each flip.
She predicted heads 47 times and got 25 correct.
Of the times she predicted tails, she got 26 correct.

a) Complete the frequency tree below to show these results.

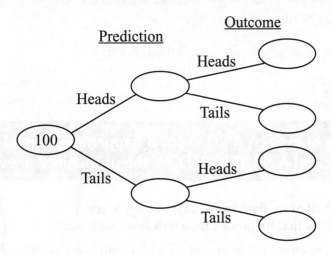

[2 marks]

b) Work out the relative frequency of Danielle predicting the outcome correctly.

....
[2 marks]

7 Mel throws a ball at a target using her left and right hands.
Her results are shown in the table on the right.

	Left Hand	Right Hand
Throws	20	100
Hit target	12	30

a) Estimate the probability that Mel will hit the
target with her next throw if she uses her left hand.

....
[2 marks]

b) Mel uses her results to estimate the probabilities of her hitting the target using
each hand. Explain which of her estimated probabilities will be more reliable.

....

....
[1 mark]

The AND / OR Rules

This page will show you two really important probability rules.

Combined Probability — Two or More Events

1) Always start by working out what different <u>SINGLE EVENTS</u> you're interested in.
2) Find the probability of <u>EACH</u> of these <u>SINGLE EVENTS</u>.
3) Apply the <u>AND/OR</u> rule.

And now for the rules. Say you have <u>two events</u> — call them A and B...

The **AND** Rule

This only works when the two events are <u>independent</u> — one event happening <u>does not affect</u> the chances of the other happening.

The probability of event A <u>AND</u> event B happening is equal to the probability of event A <u>MULTIPLIED BY</u> the probability of event B.

EXAMPLE

Dave picks one ball at random from each of bags X and Y. Find the probability that he picks a blue ball from both bags.

1) The <u>single events</u> you're interested in are 'picks a blue ball from bag X' and 'picks a blue ball from bag Y'.

2) Find the <u>probabilities</u> of the events.
   P(Dave picks a blue ball from bag X) = $\frac{4}{10}$ = 0.4

   P(Dave picks a blue ball from bag Y) = $\frac{2}{8}$ = 0.25

3) Use the <u>AND rule</u>.   P(Dave picks a blue ball from bag X <u>AND</u> bag Y) = 0.4 × 0.25 = 0.1

The **OR** Rule

This rule only works when the two events <u>can't both happen</u> at the same time.

The probability of <u>EITHER</u> event A <u>OR</u> event B happening is equal to the probability of event A <u>ADDED TO</u> the probability of event B.

EXAMPLE

A spinner with red, blue, green and yellow sections is spun — the probability of it landing on each colour is shown in the table. Find the probability of spinning either red or green.

Colour	red	blue	yellow	green
Probability	0.25	0.3	0.35	0.1

1) The <u>single events</u> you're interested in are 'lands on red' and 'lands on green'.

2) Write down the <u>probabilities</u> of the events.
   P(lands on red) = 0.25
   P(lands on green) = 0.1

3) Use the <u>OR rule</u>.   P(lands on <u>either</u> red <u>OR</u> green) = 0.25 + 0.1 = 0.35

Two rules to learn here

The way to remember this is that it's the wrong way round — you'd want AND to go with '+' but it doesn't. It's 'AND with ×' and 'OR with +'. Once you've got that, try these Exam Practice Questions.

Q1   Shaun is a car salesman. The probability that he sells a car on a Monday is 0.8.
     The probability that he sells a car on a Tuesday is 0.9.
     What is the probability that he sells a car on both Monday and Tuesday?   [2 marks]

Q2   a) What is the probability of spinning blue OR yellow on the spinner above?   [2 marks]
     b) What is the probability of spinning blue THEN green?   [2 marks]

Tree Diagrams

*Tree diagrams can really help you work out probabilities when you have a **combination of events**.*

Remember These **Four** Key **Tree Diagram Facts**

1) On any set of branches which meet at a point, the probabilities must <u>add up to 1</u>.

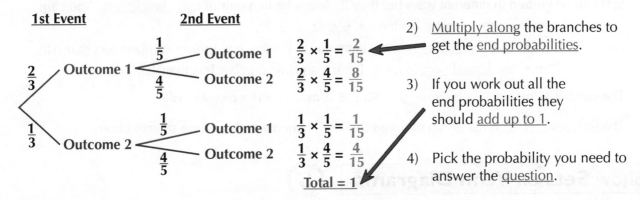

1st Event   **2nd Event**

$\frac{2}{3}$ Outcome 1

     $\frac{1}{5}$ — Outcome 1 $\frac{2}{3} \times \frac{1}{5} = \frac{2}{15}$

     $\frac{4}{5}$ — Outcome 2 $\frac{2}{3} \times \frac{4}{5} = \frac{8}{15}$

$\frac{1}{3}$ Outcome 2

     $\frac{1}{5}$ — Outcome 1 $\frac{1}{3} \times \frac{1}{5} = \frac{1}{15}$

     $\frac{4}{5}$ — Outcome 2 $\frac{1}{3} \times \frac{4}{5} = \frac{4}{15}$

Total = 1

2) <u>Multiply along</u> the branches to get the <u>end probabilities</u>.

3) If you work out all the end probabilities they should <u>add up to 1</u>.

4) Pick the probability you need to answer the <u>question</u>.

EXAMPLE

A box contains only red and green discs. A disc is taken at random and replaced. A second disc is then taken. The tree diagram below shows the probabilities of picking each colour.

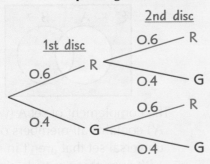

1st disc   **2nd disc**

0.6   R

    0.6 — R

    0.4 — G

0.4   G

    0.6 — R

    0.4 — G

a) **What is the probability that both discs are red?**

<u>Multiply</u> along the <u>branches</u> to find the probability you want:

P(both discs are red) = P(red, red) = 0.6 × 0.6 = 0.36

b) **What is the probability that you pick a green disc then a red disc?**

<u>Multiply</u> along the <u>branches</u> to find the probability you want:

P(green, red) = 0.4 × 0.6 = 0.24

Watch out for events that are affected by <u>what else has happened</u> — you'll get <u>different probabilities</u> on different sets of branches.

EXAMPLE

Florence either walks or drives to work. The probability that she <u>walks</u> is <u>0.3</u>. If she <u>walks</u>, the probability that she is <u>late</u> is <u>0.8</u>. If she <u>drives</u>, the probability that she is <u>late</u> is <u>0.1</u>. Complete the tree diagram below.

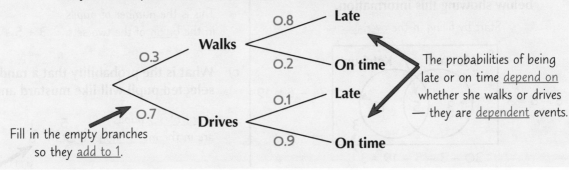

     0.8 — **Late**

Walks

0.3   0.2 — **On time**

0.7   0.1 — **Late**

Drives

     0.9 — **On time**

The probabilities of being late or on time <u>depend on</u> whether she walks or drives — they are <u>dependent</u> events.

Fill in the empty branches so they <u>add to 1</u>.

See how useful tree diagrams are

With probability questions that seem hard, drawing a tree diagram can be a good place to start.

Q1   Using the second tree diagram, find the probability of picking two green discs.   [2 marks]

Q2   Using the third tree diagram, find the probability that Florence drives to work and is late.   [2 marks]

Sets and Venn Diagrams

Venn diagrams are a way of displaying sets in intersecting circles.

A **Set** is a **Collection of Numbers** or **Objects**

1) Sets are just <u>collections of things</u> — we call these 'things' <u>elements</u>.

2) Sets can be written in different ways but they'll always be in a pair of <u>curly brackets</u> {}. You can:

   Each of these → • <u>list the elements</u> in the set, e.g. {2, 3, 5, 7}.

   describes the → • give a <u>description</u> of the elements in the set, e.g. {prime numbers less than 10}.

   same set. → • use <u>formal notation</u>, e.g. {$x : x$ is a prime number less than 10}

3) The symbol ∈ means '<u>is a member of</u>'. So $x \in$ A means 'x is a member of A'.

4) The <u>universal set</u> (ξ) is the group of things that the elements of the set are selected from.

Show **Sets** on **Venn Diagrams**

1) On a <u>Venn diagram</u>, each <u>set</u> is represented by a <u>circle</u>.
   The <u>universal set</u> is everything <u>inside</u> the <u>rectangle</u>.

2) The diagram can show either the <u>actual elements</u> of each set, or the <u>number of elements</u> in each set.

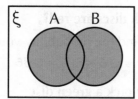

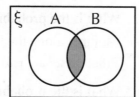

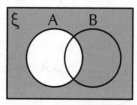

The <u>union</u> of sets A and B (written A ∪ B) contains all the elements in <u>either</u> set A <u>or</u> set B — it's everything <u>inside the circles</u>.

The <u>intersection</u> of sets A and B (written A ∩ B) contains all the elements in <u>both</u> set A <u>and</u> set B — it's where the <u>circles overlap</u>.

The <u>complement</u> of set A (written A') contains all members of the universal set that <u>aren't</u> in set A — it's everything <u>outside circle A</u>.

EXAMPLE   In a class of 30 pupils, 8 of them like mustard, 24 of them like ketchup and 5 of them like both mustard and ketchup.

a) **Complete the Venn diagram below showing this information.**

   Start by filling in the <u>overlap</u>.

   8 − 5 = 3 ──→ ③ ⑤ ⑲ ←── 24 − 5 = 19
   ξ Mustard / Ketchup
   3
   30 − 3 − 5 − 19 = 3

b) **How many pupils like mustard or ketchup?**

   This is the number of pupils in the <u>union</u> of the two sets.   3 + 5 + 19 = 27

c) **What is the probability that a randomly selected pupil will like mustard and ketchup?**

   <u>5 out of 30</u> pupils are in the <u>intersection</u>. $\frac{5}{30} = \frac{1}{6}$

   ⌇ This is P(M ∩ K). ⌇

Learn what each bit of a Venn diagram represents

Once you've got your head round the new words to learn, test your skills with this Exam Question.

Q1 50 birdwatchers were looking for pigeons and seagulls. 28 of them saw a pigeon, 15 saw both birds and 10 didn't see either bird. Show this information on a Venn diagram and use it to find the probability that a randomly selected birdwatcher saw a seagull.   [4 marks]

QI Video Solution

Warm-up and Worked Exam Questions

Probability can be tricky to get your head around so it's important to get loads of practice.
Try these warm-up questions and take a look back at anything you're unsure about.

Warm-up Questions

1) What is the probability of rolling a six three times in a row with a six-sided dice?

2) Dmitri is practising front flips. The probability that he lands correctly on his first try is 0.5.
   The probability that he lands correctly on his second try is 0.7.
   What is the probability that he lands correctly on both his first and second tries?

3) For bags X and Y on page 160, find the probability of picking:
   a) an orange ball from both bags
   b) an orange ball from bag X OR bag Y.

4) A bag contains 6 red balls and 4 black ones. If one ball is picked at random, placed back into the bag, then another ball is drawn at random, find the probability that they're both red.

5) There are 40 people at a pet owners convention. 13 of them own a cat,
   21 of them own a dog and 5 of them own a cat and a dog.
   a) Show this information on a Venn diagram.
   b) Find the probability that a randomly selected pet owner owns a cat but not a dog.

Worked Exam Question

Look through this worked exam question and make sure you understand it — don't leave it to chance.

1   A couple are both carriers of a gene that causes a disease.
    If they have a child, the probability that the child will carry the gene is 0.25.

   a) The couple have two children. Draw a tree diagram to show the probabilities of each child carrying or not carrying the gene.

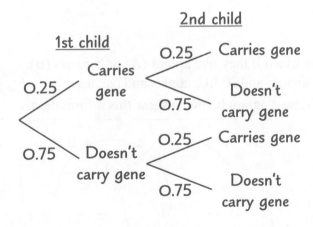

[1 mark]

   b) Find the probability that both children will carry the gene.

   P(both children carry the gene) = 0.25 × 0.25 = 0.0625

   0.0625
   [2 marks]

Exam Questions

2   A biased 5-sided spinner is numbered 1-5.
     The probability that the spinner will land on each of the numbers 1 to 5 is given in this table.

Number	1	2	3	4	5
Probability	0.3	0.15	0.2	0.25	0.1

   a)   What is the probability of the spinner landing on a 4 or a 5?

   ....
   [2 marks]

   b)   The spinner is spun twice. What is the probability that it will land
        on a 1 on the first spin and a 3 on the second spin?

   ....
   [2 marks]

3   Chelsea is playing the game 'hook-a-duck'. The probability that she wins a prize is 0.3.

   a)   What is the probability that she does not win a prize?

   ....
   [1 mark]

   b)   If she plays two games, what is the probability that she doesn't win a prize in either game?

   ....
   [2 marks]

4   100 Year 7 students were asked if they like apples (A) or bananas (B).
     70 like apples, 40 like bananas and 20 like apples and bananas.

   a)   Complete the Venn diagram below to show this information.

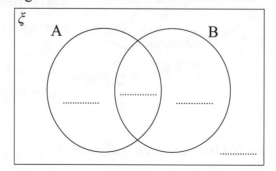

   It's a good idea to
   start by filling in
   the intersection.

   [3 marks]

   b)   One of the students is selected at random.
        Find $P(A \cup B)$.

   ....
   [2 marks]

Sampling and Bias

Sampling is about using what you know about **smaller** groups to tell you about **bigger** groups.

Use a **Sample** to **Find Out About** a Population

1) For any statistical project, you need to find out about a group of people or things. E.g. all the pupils in a school, or all the trees in a forest. This whole group is called the POPULATION.

2) Often you can't collect information about every member of the population because there are too many. So you select a smaller group from the population, called a SAMPLE, instead.

3) It's really important that your sample fairly represents the WHOLE population. This allows you to apply any conclusions about your sample to the whole population.

E.g. if you find that ¾ of the people in your sample like cheese, you can estimate that ¾ of the people in the whole population like cheese.

You Need to **Spot Problems** with **Sampling Methods**

A BIASED sample (or survey) is one that doesn't properly represent the whole population.

To SPOT BIAS, you need to think about:

1) **WHEN**, **WHERE** and **HOW** the sample is taken.
2) **HOW MANY** members are in it.

If certain groups are left out of the sample, there can be BIAS in things like age, gender, or different interests. If the sample is too small, it's also likely to be biased.

Bigger populations need bigger samples to represent them.

EXAMPLE

Samir's school has 800 pupils. Samir is interested in whether these pupils would like to have more music lessons. For his sample he selects 5 members of the school orchestra to ask.
Explain why the opinions Samir collects from his sample might not represent the whole school.

Only members of the orchestra are included, so the opinions are likely to be biased in favour of more music lessons. And a sample of 5 is too small to represent the whole school.

If possible, the best way to AVOID BIAS is to select a large sample at random from the whole population.

Simple Random Sampling — choosing at **Random**

One way to get a random sample is to use 'simple random sampling'.

To SELECT a SIMPLE RANDOM SAMPLE...

1) Assign a number to every member of the population.
2) Create a list of random numbers.
3) Match the random numbers to members of the population

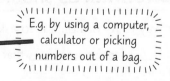

E.g. by using a computer, calculator or picking numbers out of a bag.

Before you begin collecting data, think about your sampling method

You want to make sure that you choose a sample that represents your population. Make sure you know how to spot poor sampling methods too. Then you'll be ready to take on this Exam Practice Question.

Q1 Tina wants to find out how often people in the UK travel by train. She decides to ask 20 people waiting for trains at her local train station one morning. Comment on whether Tina can use the results of her survey to draw conclusions about the whole population. [2 marks]

Collecting Data

*Data you **collect yourself** is called **primary** data. If you use data that **someone else has collected**, e.g. you get it from a website, it's called **secondary** data. You need to **record** primary data in a way that's **easy to analyse** and **suitable** for the **type** of data you've got.*

There are **Different Types** of Data

QUALITATIVE DATA is <u>descriptive</u>. It uses <u>words</u>, not numbers.	E.g. <u>pets' names</u> — Smudge, Snowy, Dave, etc. <u>Favourite flavours of ice cream</u> — 'vanilla', 'caramel-marshmallow-ripple', etc.
QUANTITATIVE DATA measures <u>quantities</u> using <u>numbers</u>.	E.g. <u>heights</u> of people, <u>times taken</u> to finish a race, <u>numbers of goals</u> scored in football matches, and so on.

There are two types of <u>quantitative</u> data.

DISCRETE DATA
1) It's <u>discrete</u> if the numbers can only take certain <u>exact</u> values.
2) E.g. the number of customers in a shop each day has to be a whole number — you can't have half a person.

CONTINUOUS DATA
1) If the numbers can take <u>any value</u> in a range, it's called <u>continuous</u> data.
2) E.g. heights and weights are continuous measurements.

You can **Organise** your **Data** into **Classes**

1) To record data in a <u>table</u>, you often need to <u>group</u> it into <u>classes</u> to make it more manageable. <u>Discrete</u> data classes should have '<u>gaps</u>' between them, e.g. '<u>0-1 goals</u>', '<u>2-3 goals</u>' (it jumps from 1 to 2 because there are no values in between). <u>Continuous</u> data classes should have <u>no 'gaps'</u>, so are often written using <u>inequalities</u> (see p.177).

2) Whatever the data you have, make sure <u>none of the classes overlap</u> and that they <u>cover all the possible values</u>.

When you <u>group</u> data you <u>lose</u> <u>some accuracy</u> because you don't know the exact values any more.

EXAMPLE

Jonty wants to find out about the ages (in whole years) of people who use his local library. Design a table he could use to collect his data.

Include <u>columns</u> for: the <u>data values</u>, '<u>Tally</u>' to count the data and '<u>Frequency</u>' to show the totals.

Use <u>non-overlapping</u> classes — with <u>gaps</u> because the data's <u>discrete</u>.

Include classes like '<u>...or over</u>', '<u>...or less</u>' or '<u>other</u>' to <u>cover all options</u> in a sensible number of classes.

Age (whole years)	Tally	Frequency
0-19		
20-39		
40-59		
60-79		
80 or over		

Questionnaires should be **Designed Carefully**

Another way to record data is to ask people to fill in a <u>questionnaire</u>. Your <u>questions</u> should be:

<u>Watch out</u> for <u>response boxes</u> that could be <u>interpreted</u> in <u>different</u> ways, that <u>overlap</u>, or that <u>don't</u> <u>allow</u> for <u>all</u> possible answers.

1) **Clear** and **easy to understand**
2) **Easy** to **answer**
3) **Fair** — not leading or biased

A question is '<u>leading</u>' if it guides you to picking a particular answer.

Tables are a really good way to record data

You need to know what type of data you've got so you can record and display it in a suitable way.

Q1   Penny asks some students how many times they went to the cinema in the last year. Say whether this data is discrete or continuous and design a table to record it in.   [2 marks]

Mean, Median, Mode and Range

Mean, median, mode and range pop up all the time in exams — make sure you know what they are.

MODE = MOST common

MEDIAN = MIDDLE value (when values are in order of size)

MEAN = TOTAL of items ÷ NUMBER of items

RANGE = Difference between highest and lowest

REMEMBER:

Mode = most (emphasise the 'mo' in each when you say them)

Median = mid (emphasise the m*d in each when you say them)

Mean is just the average, but it's mean 'cos you have to work it out.

There's one <u>Golden Rule</u> for finding the <u>median</u> that lots of people forget:

Always REARRANGE the data in ASCENDING ORDER (and check you have the same number of entries!)

You <u>must</u> do this when finding the median, but it's also really useful for working out the <u>mode</u> too.

EXAMPLE   **Find the <u>median</u>, <u>mode</u>, <u>mean</u>, and <u>range</u> of these numbers: 2, 5, 3, 2, 0, 1, 3, 3**

1   <u>Rearrange</u> the numbers into ascending order. The **MEDIAN** is the <u>middle value</u>.

When there are <u>two middle numbers</u>, the median is <u>halfway</u> between the two.

0, 1, 2, (2, 3) 3, 3, 5
← 4 numbers either side →

Median = 2.5

To find the <u>position</u> of the median of n values, you can use the formula (n + 1) ÷ 2. Here, (8 + 1) ÷ 2 = position 4.5 — that's halfway between the 4th and 5th values.

2   <u>MODE</u> (or <u>modal value</u>) is the most common value = 3

3 $\underline{\text{MEAN}} = \dfrac{\text{total of items}}{\text{number of items}} = \dfrac{0 + 1 + 2 + 2 + 3 + 3 + 3 + 5}{8} = \dfrac{19}{8} = 2.375$

4   <u>RANGE</u> = difference between highest and lowest values = 5 − 0 = 5

A Trickier Example

EXAMPLE   **The heights (to the nearest cm) of 8 penguins at a zoo are 41, 43, 44, 44, 47, 48, 50 and 51. Two of the penguins are moved to a different zoo. If the mean height of the remaining penguins is 44.5 cm, find the heights of the two penguins that moved.**

$\text{Mean} = \dfrac{\text{total height}}{\text{no. of penguins}}$

So <u>total height = no. of penguins × mean</u>

Total height of 8 penguins = 368 cm.
Total height of remaining 6 penguins = 6 × 44.5 = 267 cm.
Combined height of penguins that moved = 368 − 267 = 101 cm.
So the heights must be 50 cm and 51 cm.

Mean, median, mode, range — easy marks for learning 4 definitions

The maths involved in working these out is simple, so you'd be mad not to learn the definitions.

Q1   Find the mean, median, mode and range for these test scores:
     6, 15, 12, 12, 11   [4 marks]

Q2   A set of 8 heights has a mean of 1.6 m. A new height of 1.5 m is added.
     Explain whether the mean of all 9 heights will be higher or lower than 1.6 m. [1 mark]

Q1 Video Solution

Exam Questions

2   A survey was carried out in a local cinema to find out which flavour of popcorn people bought. The results are in the table below.

a)   Draw and label a pie chart to represent the information.

Type of popcorn	Number sold
Plain	12
Salted	18
Sugared	9
Toffee	21

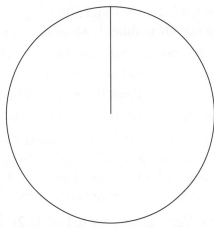

[4 marks]

Another survey was carried out to find out which flavour of ice cream people bought. The results are shown in the pie chart on the right.

Kim compares the two pie charts and says, "The results show that more people chose strawberry ice cream than toffee popcorn."

b)   Explain whether or not Kim is right.

....

....

....

[1 mark]

3   This data shows the amount of rainfall in mm that 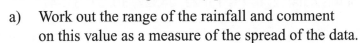 fell on an island during a 12-day period in June.

a)   Work out the range of the rainfall and comment on this value as a measure of the spread of the data.

Don't be put off by the way the data is displayed — look at the key to work out how to read off the values.

```
0 | 8
1 | 7 9
2 | 3 6 9
3 | 0 1 4 7 8
6 | 3
```

Key
0 | 8 means 8 mm of rain

....

....

[3 marks]

In November the median amount of rainfall was 22 mm and the range was 20 mm.

b)   Compare the rainfall in June with the rainfall in November.

....

....

....

[3 marks]

Revision Questions for Section Seven

Here's the inevitable list of questions to test how much you know.
- Tick off each question when you've got it right.
- When you're completely happy with a topic, tick that off too.

For even more practice, try the Sudden Fail Quiz for Section Seven — just scan this QR code!

Section Seven Quiz

Probability (p153-156) ☑

1) I pick a random number between 1 and 50.
Find the probability that my number is a multiple of 6. ☑

2) The probability of a spinner landing on red is 0.3. What is the probability it doesn't land on red? ☑

3) I flip a fair coin 3 times. a) Using H for heads and T for tails list all the possible outcomes.
b) What is the probability of getting exactly one head? ☑

4) What are the formulas for: a) relative frequency? b) expected frequency? ☑

5) 160 people took a 2-part test. 105 people passed the first part and of these,
60 people passed the second part. 25 people didn't pass either part.
a) Show this information on a frequency tree.   b) Find the relative frequency of each outcome.
c) If 300 more people do the test, estimate how many of them would pass both parts. ☑

Harder Probability and Venn Diagrams (p160-162) ☑

6) The table shows the probabilities of a biased dice landing on each number. What is the probability of it landing on 1 or 4?

Number	1	2	3	4	5	6
Probability	0.2	0.15	0.1	0.3	0.15	0.1
☑

7) I have a standard pack of 52 playing cards. Use a tree diagram to find the probability of
me picking two cards at random and getting no hearts if the first card is replaced. ☑

8) 100 people were asked whether they like tea or coffee. Half the people said they like coffee,
34 people said they like tea, 20 people said they like both.
a) Show this information on a Venn diagram.
b) If one of the 100 people is randomly chosen, find the probability of them liking tea or coffee. ☑

Collecting Data and Finding Averages (p165-167) ☑

9) What is a sample and why does it need to be representative? ☑

10) Is 'eye colour' qualitative, discrete or continuous data? ☑

11) Complete this frequency table for the data below. ➞

Pet	Tally	Frequency

Cat, Cat, Dog, Dog, Dog, Rabbit, Fish, Cat, Rabbit, Rabbit, Dog, Dog, Cat, Cat, Dog, Rabbit, Cat, Fish, Cat, Cat ☑

12) Find the mode, median, mean and range of this data: 2, 8, 11, 15, 22, 24, 27, 30, 31, 31, 41 ☑

Graphs and Charts (p170-173) ☑

13) How do you find frequencies from a pictogram? ☑

14) The table opposite shows how some students rated a film.

Film rating	Terrible	Bad	OK	Good	Amazing
Students	40	30	40	45	25

Draw a suitable diagram to show the number of students giving each rating. ☑

15) a) Draw a line graph to show the time series data in this table.
b) Describe the repeating pattern in the data.

Quarter	1	2	3	4	1	2	3	4
Sales (1000's)	1	1.5	1.7	3	0.7	0.9	1.2	2.2
☑

16) Use data from Q14 to draw a suitable diagram showing the proportion of students giving each rating. ☑

17) Sketch graphs to show: a) weak positive correlation, b) strong negative correlation, c) no correlation ☑

Frequency Tables and Averages (p176-177) ☑

18) For this grouped frequency table showing the lengths of some pet alligators:
a) find the modal class,   b) find the class containing the median,
c) estimate the mean.

Length (y, in m)	Frequency
$1.4 \le y < 1.5$	4
$1.5 \le y < 1.6$	8
$1.6 \le y < 1.7$	5
$1.7 \le y < 1.8$	2
☑

Interpreting and Comparing Data Sets (p178-179) ☑

19) Explain the effect that outliers can have on the mean and range of data. ☑

20) These pie charts show the results of a survey on the colour of people's cars.
Compare the popularity of each colour of car amongst men and women.

Men: Red, Green, Black, Blue
Women: Silver, Green, Red, Blue
☑

Practice Paper 1: Non-calculator
As final preparation for the exams, we've included three full practice papers to really put your Maths skills to the test. Paper 1 is a non-calculator paper — Paper 2 and Paper 3 (on pages 196 and 209) require a calculator. There's a whole page on formulas on p.244. Good luck...

Candidate Surname		Candidate Forename(s)	
Centre Number	**Candidate Number**	**Candidate Signature**	

GCSE

Mathematics   **Foundation Tier**

Paper 1 (Non-Calculator)

Practice Paper
Time allowed: 1 hour 30 minutes

You must have:
Pen, pencil, eraser, ruler, protractor, pair of compasses.
You may use tracing paper.

You are **not allowed** to use a calculator.

Instructions to candidates
*   Use **black** ink to write your answers.
*   Write your name and other details in the spaces provided above.
*   Answer **all** questions in the spaces provided.
*   In calculations, show clearly how you worked out your answers.
*   Do all rough work on the paper.

Information for candidates
*   The marks available are given in brackets at the end of each question.
*   You may get marks for method, even if your answer is incorrect.
*   There are 26 questions in this paper. There are no blank pages.
*   There are 80 marks available for this paper.

Answer ALL the questions.

Write your answers in the spaces provided.

You must show all of your working.

1   Write 0.113 as a fraction.
Circle your answer.

$\dfrac{113}{100}$ $\qquad\qquad$ $\dfrac{113}{10\ 000}$ $\qquad\qquad$ $\dfrac{113}{1000}$ $\qquad\qquad$ $\dfrac{13}{100}$

[Total 1 mark]

2   Write the ratio 40 : 25 in its simplest form.

....

[Total 1 mark]

3   Eight points are shown plotted on the grid.

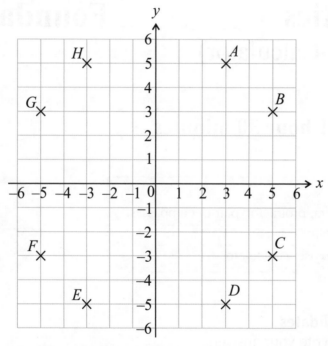

(a)   Circle the point that has coordinates (–5, –3).

$\qquad$ C $\qquad\qquad\qquad$ E $\qquad\qquad\qquad$ F $\qquad\qquad\qquad$ G

[1]

(b)   Circle the equation of the straight line that passes through points *A* and *D*.

$\qquad x = 3 \qquad\qquad x + y = 3 \qquad\qquad y = 3x \qquad\qquad y = 3$

[1]

[Total 2 marks]

4 Convert 3.97 km into m.

.... m

[Total 1 mark]

---

5 Karl has five number cards.

$$\boxed{-6} \quad \boxed{6} \quad \boxed{-8} \quad \boxed{-12} \quad \boxed{2}$$

(a) Write Karl's number cards in order, starting with the lowest.

lowest   .... , ... , ... , ... , ...   highest

[1]

(b) Use two of Karl's number cards to make this calculation correct.

....   −   ....   = 10

[1]

[Total 2 marks]

---

6 Beth has some 5p and 10p coins.

Coin	Number
5p	28
10p	41

She changes her coins for 50p coins at the bank.

How many 50p coins does she receive?

....

[Total 3 marks]

---

7 Calculate

$$\frac{1.2 - 0.2 \times 4}{0.05}$$

....

[Total 2 marks]

8

> If you add a multiple of 3 to a multiple of 6, you always get a multiple of 9.

Give an example to show that this statement is not true.

....

....

[Total 1 mark]

---

9 (a) Simplify $11a + 5b - 2a + 2b$

Circle your answer.

$$13a + 3b \qquad 9a + 7b \qquad 13a + 7b \qquad 16ab$$

[1]

(b) Simplify $2a \times 3a$

Circle your answer.

$$5a \qquad 6a \qquad 5a^2 \qquad 6a^2$$

[1]

[Total 2 marks]

---

10 Chloe invests £300 in a bank account.
The account pays 2% simple interest each year.

Work out how much money she has in her account after 4 years.

£ ....

[Total 3 marks]

---

11 The dual bar chart below shows the favourite sports of the pupils in a class.

One bar is missing.

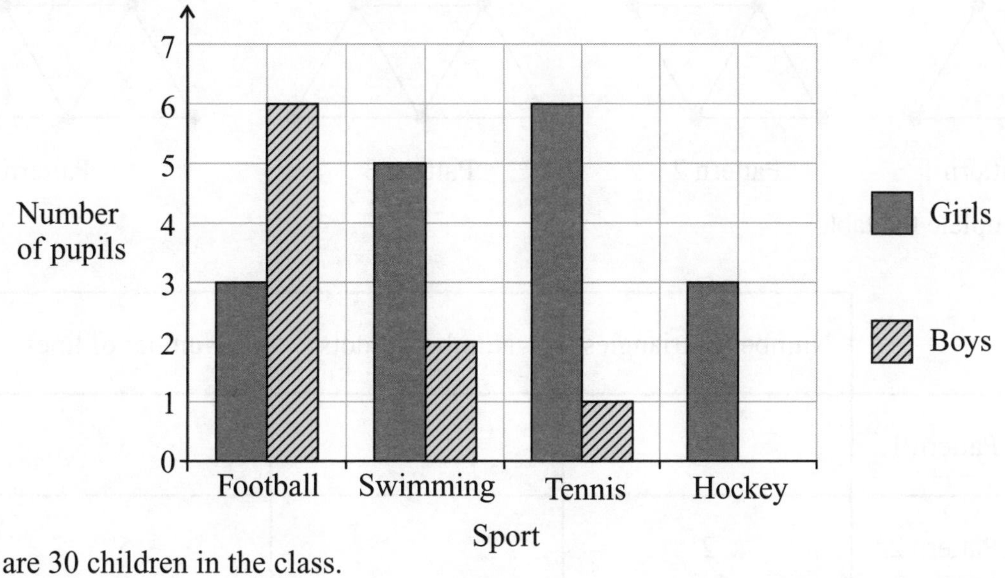

There are 30 children in the class.

(a) Draw the missing bar to show the number of boys whose favourite sport is Hockey.

[2]

(b) One child is chosen at random from the class.
Find the probability that their favourite sport is swimming.

....

[2]

(c) What is the ratio of the number of boys who chose swimming to the number of girls who chose tennis? Give your answer in its simplest form.

....

[2]

[Total 6 marks]

12 Decide whether the sequence is an arithmetic or geometric progression, and write down in words the rule to get from one term to the next.

2, 8, 32, 128...

Arithmetic ☐   Geometric ☐

Rule: ....

....

[Total 2 marks]

4

13 The diagram shows the first four patterns in a sequence.

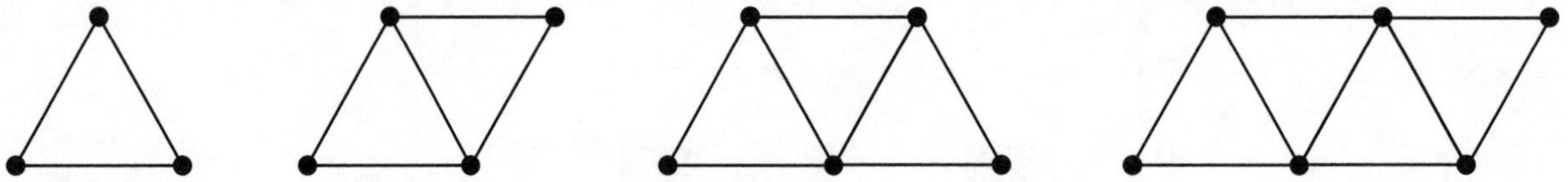

Pattern 1   Pattern 2   Pattern 3   Pattern 4

(a) Complete the table.

	Number of triangles	Number of dots	Number of lines
Pattern 1	1	3	3
Pattern 2	2	....	5
Pattern 3	....	5	....
Pattern 4	4	....	....

[1]

(b) Work out the number of lines in pattern 10.

....

[2]

(c) (i) Find a formula for the number of dots D in Pattern n.

....

[2]

(ii) Find the number of dots in pattern 200.

....

[1]

[Total 6 marks]

14 A theatre sells three types of tickets.

Ticket type	Cost
Adult	£9
Child	£5
Senior	£6.50

The pictogram shows the number of tickets of each type sold for one performance.

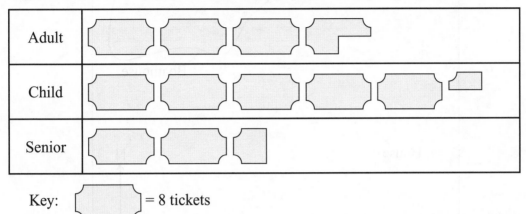

Key: [ticket] = 8 tickets

How much money did the theatre make from all the ticket sales for this performance?

£ ....

[Total 6 marks]

6

15 The scale drawing shows the gardens of a country house.

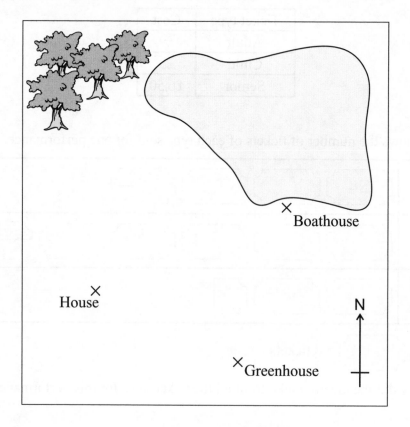

Scale: 1 cm = 100 metres

(a) Find the three-figure bearing of the boathouse from the house.

.................°

[1]

(b) Find the actual distance from the boathouse to the greenhouse.

.... metres

[2]

(c) A summerhouse is

450 metres from the house
700 metres from the greenhouse

Plot with a cross (×) the position of the summerhouse on the map.
Do not rub out your construction lines.

[2]

[Total 5 marks]

16 Angie makes wedding cakes with three tiers.

She needs 800 grams of sultanas to make the bottom tier of a cake.
The middle tier needs 75% of the ingredients required for the bottom tier.
The top tier needs 50% of the ingredients of the bottom tier.

Angie needs to make five wedding cakes. She has 8 kilograms of sultanas.

Does Angie have enough sultanas to make five cakes? Show your working.

[Total 5 marks]

17 Carrots cost 69p per kilogram.
Ahmed buys 2.785 kilograms of carrots.

(a) Estimate the cost, in pence, of his carrots.
Show the numbers you use to work out your estimate.

.... p
[2]

(b) Is your estimate in (a) bigger or smaller than the actual cost?
Tick the correct answer.

Bigger ☐   Smaller ☐

Explain your answer.

....

....
[1]

[Total 3 marks]

18 The graph can be used to convert between pounds (£) and Australian dollars ($).

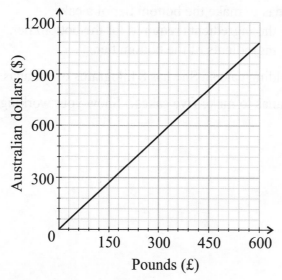

Jack goes on holiday to Australia and China.

(a) He changes £300 into Australian dollars ($).
How many Australian dollars does he get for £300?

$ ....
[1]

(b) He spends $390 of his money while in Australia.
He converts the rest of his Australian dollars into Chinese yuan using the exchange rate

1 Australian dollar = 6 Chinese yuan

How many Chinese yuan does Jack get?

.... yuan
[2]

[Total 3 marks]

19 Work out the value of *k* if

$$k \times 3^{-2} = 4$$

k = ...
[Total 2 marks]

20 The diagram shows an isosceles triangle.

Three of these isosceles triangles fit together with three squares around a point *O*.

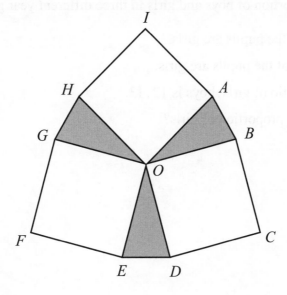

Show clearly that angle *OAB* = 75°.

21 Work out $1\frac{2}{3} \times 1\frac{5}{8}$. Give your answer as a mixed number.

....

22 Write 594 000 000 000 in standard form.

23 A school records the proportion of boys and girls in three different year groups.

In Year 9, $\frac{9}{20}$ of the pupils are girls.

In Year 10, 49% of the pupils are girls.

In Year 11, the ratio of girls : boys is 12 : 13.

Which year has the largest proportion of girls?

[Total 3 marks]

24 The diagram shows a circle *A* and a sector *B*.

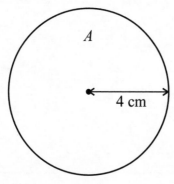

Not to scale

Show that the area of *A* is twice the area of *B*.

[Total 4 marks]

25 Quadrilateral *ABCD* is made up of two right-angled triangles, *ABC* and *ACD*.

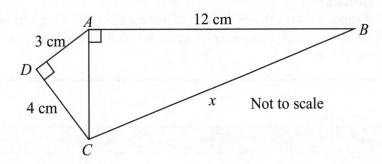

Not to scale

(a) Find the length of the side labelled *x*.

.... cm

[4]

(b) Find the area of quadrilateral *ABCD*.

.... cm²

[2]

[Total 6 marks]

26 Solve the simultaneous equations

$$3x + 2y = 17$$
$$2x + y = 10$$

x = ...

y = ...

[Total 3 marks]

[TOTAL FOR PAPER = 80 MARKS]

12

Practice Paper 2: Calculator
Right, here's Practice Paper 2 — you'll need a calculator for this one. Don't forget there's a page on formulas (p.244) — it tells you which formulas you'll be given in your exam and which you'll need to learn.

Candidate Surname		Candidate Forename(s)

Centre Number	Candidate Number	Candidate Signature

GCSE

Mathematics
Paper 2 (Calculator)

Foundation Tier

Practice Paper
Time allowed: 1 hour 30 minutes

You must have:
Pen, pencil, eraser, ruler, protractor, pair of compasses.
You may use tracing paper.

You **may use** a calculator.

Instructions to candidates
- Use **black** ink to write your answers.
- Write your name and other details in the spaces provided above.
- Answer **all** questions in the spaces provided.
- In calculations, show clearly how you worked out your answers.
- Do all rough work on the paper.
- Unless a question tells you otherwise, take the value of π to be 3.142, or use the π button on your calculator.

Information for candidates
- The marks available are given in brackets at the end of each question.
- You may get marks for method, even if your answer is incorrect.
- There are 28 questions in this paper. There are no blank pages.
- There are 80 marks available for this paper.

Answer ALL the questions.

Write your answers in the spaces provided.

You must show all of your working.

1   Write $\frac{3}{5}$ as a percentage.
    Circle your answer.

    6%   30%   15%   60%

[Total 1 mark]

2   A function is represented by this number machine.

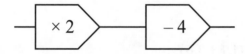

    The **output** of the machine is 20. Circle the input.

    8   12   14   36

[Total 1 mark]

3   Complete this bill.

<table>
<tr><td colspan="4" align="center">**Barbara's Café**</td></tr>
<tr><td>**Menu Item**</td><td>**Number Ordered**</td><td>**Cost per Item**</td><td>**Total**</td></tr>
<tr><td>Tea</td><td>2</td><td>£1.25</td><td>£2.50</td></tr>
<tr><td>Coffee</td><td>........</td><td>£1.60</td><td>£9.60</td></tr>
<tr><td>Cake</td><td>4</td><td>£...............</td><td>£5.20</td></tr>
<tr><td>Tip</td><td></td><td></td><td>£2.50</td></tr>
<tr><td></td><td></td><td>**Total cost**</td><td>£...............</td></tr>
</table>

[Total 3 marks]

1

4 (a) Draw a line to match each shape to its number of **surfaces**.

cone   4

                        3

sphere   2

cylinder   1

[2]

(b) Write down the number of **vertices** for a triangular prism.

....

[1]

[Total 3 marks]

5 (a) Calculate

$$\sqrt{12.2} + (1.1 + 3.6)^3$$

Write down all the digits on your calculator.

....

[1]

(b) Round your answer to (a) to two decimal places.

....

[1]

[Total 2 marks]

6 Kamil has these four number cards.

3   5   4   6

List the eight even numbers greater than 4000 Kamil can make by rearranging all four cards.

[Total 2 marks]

7 Circle the number below that is both a square number and a cube number.

16 27 64 25 8 100

[Total 1 mark]

8 (a) Expand

$$4(a + 2)$$

....

[1]

(b) Factorise

$$y^2 + 5y$$

....

[1]

[Total 2 marks]

9 The ages (in years) of seven children are

6 12 9 6 5 7 11

(a) Find the mode of the ages.

....

[1]

(b) Find the median age.

....

[1]

(c) Find the mean age.

....

[2]

[Total 4 marks]

10 An equilateral triangle *T* is shown on the grid.

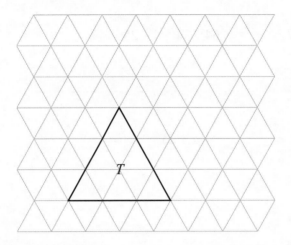

(a) Another triangle congruent to *T* is joined to *T* to form a quadrilateral.
Write down the number of lines of symmetry of the quadrilateral.

....

[1]

(b) Show on the grid how four triangles congruent to *T* can be joined together
to form a shape with rotational symmetry of order 3.

[1]

[Total 2 marks]

11 Work out 185% of £3500.

£ ....

[Total 2 marks]

12 Sandra attends a job interview at a school.
The school refunds her travelling expenses if she uses the cheapest possible method of transport.

There are two methods of transport that Sandra can use to attend the interview.

Method 1: By car	**Method 2: By car and train**
Sandra lives 27 miles from the school.	Sandra lives 4 miles from the station. The cost of a return train ticket is £17.60.

The school refunds car travel at a rate of 40p per mile.

Which method should Sandra use to travel to her interview and home again if she wants a refund for her expenses?

Show how you work out your answer.

[Total 3 marks]

13 A car park contains 28 cars and 16 motorbikes.

$\frac{3}{4}$ of the cars and $\frac{3}{8}$ of the motorbikes are red.

A red vehicle is chosen at random.

What is the probability that it is a car? Give your answer as a fraction in its simplest form.

....

[Total 3 marks]

14 (a) Complete the table of values for $y = 7 - 2x$.

x	–2	–1	0	1	2	3	4
y		9	7			1	

[2]

(b) Draw the graph of $y = 7 - 2x$ for values of x between –2 and 4.

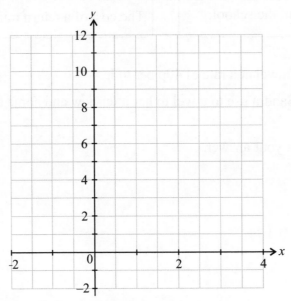

[2]

(c) What is the gradient of the line $y = 7 - 2x$?
Circle your answer.

$$-2 \qquad -1 \qquad 2 \qquad 7$$

[1]

[Total 5 marks]

15 Solve the equation $3(2x - 4) = 2x + 8$

$x =$ ....

[Total 3 marks]

16 Kieron works out $5 \times \dfrac{2}{3}$ and gets the answer $\dfrac{10}{15}$.

Explain what mistake Kieron has made in calculating his answer.

....

....

[Total 1 mark]

17 The budget airline 'Fly By Us' produces this graph to show how their passenger numbers have increased.

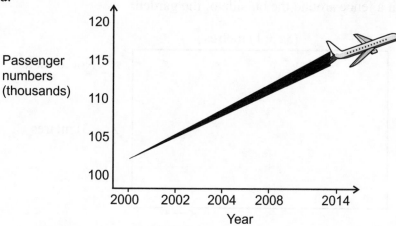

Give three different criticisms of the graph.

Criticism 1 ....

....

Criticism 2 ....

....

Criticism 3 ....

....

[Total 3 marks]

18 *AC* and *DG* are straight lines.

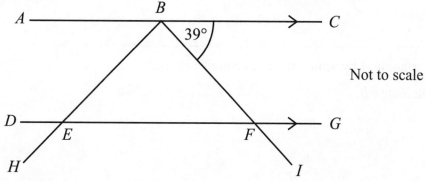

Not to scale

BH is perpendicular to *BI*.

Work out the size of angle *DEH*.
Show how you work out your answer.

.........................°

[Total 3 marks]

7

19 Jimmy has a rectangular vegetable garden measuring $(3x + 1)$ metres by $(2x - 3)$ metres.

Jimmy wants to put a fence around the outside of the garden.

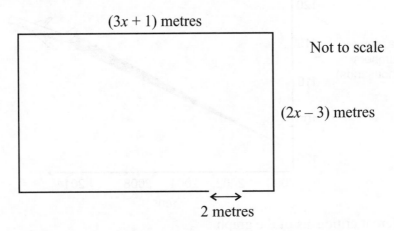

$(3x + 1)$ metres

Not to scale

$(2x - 3)$ metres

2 metres

He needs a 2 metre gap along one edge so that he can get in and out.

(a) Show that the length of the fence, L m, is given by the formula $L = 10x - 6$.

[2]

(b) Show that L is always an even number when x is a whole number.

[2]

[Total 4 marks]

20 Enlarge triangle A by scale factor 3, centre P.

Label the image B.

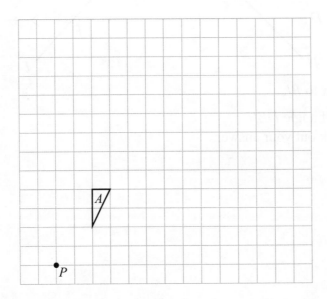

[Total 2 marks]

8

21 $\xi = \{1, 2, 3, \ldots, 10\}$
$A = \{x : 2 < x \le 6\}$
$B = \{x : x \text{ is a factor of } 12\}$

Complete the Venn diagram to show the elements of each set.

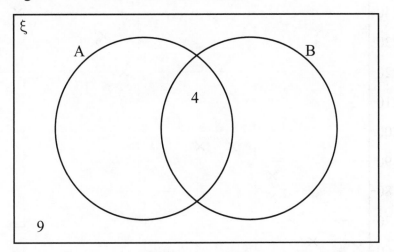

22 Orange juice and lemonade are mixed in the ratio $3 : 5$ to make orangeade.

Orange juice costs £1.60 per litre.
Lemonade costs £1.20 per litre.

What is the cost of making 18 litres of orangeade?

£ ....

[Total 4 marks]

23 Make x the subject of the formula

$$y = \frac{x^2 - 2}{5}$$

....

[Total 2 marks]

24 The scatter graph shows the maximum power (in kW)
    and the maximum speed (in km/h) of a sample of cars.

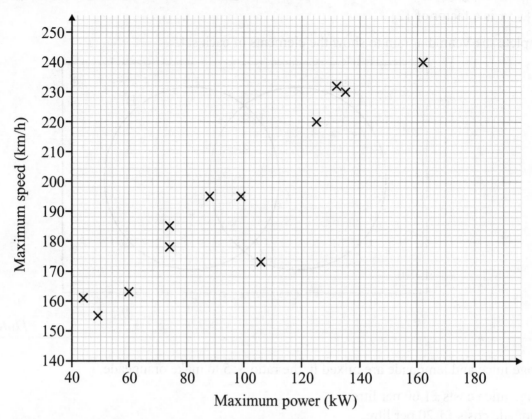

(a) One of the cars has a maximum speed of 220 km/h.
    Write down the maximum power of this car.

.... kW

[1]

(b) One of the points is an outlier as it does not fit in with the trend.
    Draw a ring around this point on the graph.

[1]

(c) Ignoring the outlier, describe the correlation shown on the scatter graph.

.... correlation

[1]

(d) A different car has a maximum power of 104 kW.
    By drawing a suitable line on your scatter graph, estimate the maximum speed of this car.

.... km/h

[2]

(e) Explain why it may not be reliable to use the scatter graph to estimate
    the maximum speed of a car with a maximum power of 190 kW.

....

....

[1]

[Total 6 marks]

25 Ollie and Amie each have an expression.

Ollie
$(x + 4)^2 - 1$ Amie
$(x + 5)(x + 3)$

Show clearly that Ollie's expression is equivalent to Amie's expression.

[Total 3 marks]

26 A company consists of 80 office assistants and a number of managers.

The pie chart shows how the 80 office assistants travel to work.

Office Assistants

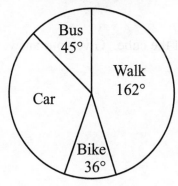

(a) How many office assistants travel to work by car?

....

[2]

18 of the managers travel by car.
Overall, 40% of the people in the company travel by car.

(b) Work out how many people there are in the company.

....

[2]

[Total 4 marks]

11

27 The diagram shows a solid aluminium cylinder and a solid silver cube.

Cylinder (aluminium)   Cube (silver)

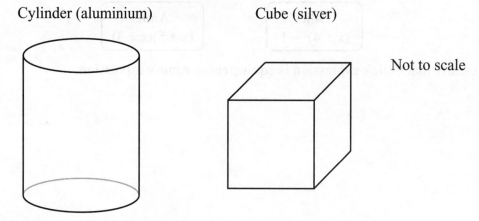

Not to scale

- The volume of the cylinder is 1180 cm³.
- The cylinder and the cube have the same mass.
- The density of aluminium is 2.7 g/cm³ and the density of silver is 10.5 g/cm³.

(a) Calculate the mass of the cylinder.

.... g

[2]

(b) Calculate the side length of the cube. Give your answer correct to two significant figures.

.... cm

[4]

[Total 6 marks]

28 The diagram shows a right-angled triangle.

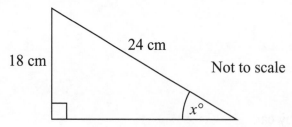

18 cm

24 cm

Not to scale

$x°$

Calculate the value of x. Give your answer correct to 1 decimal place.

$x =$ ....

[Total 2 marks]

[TOTAL FOR PAPER = 80 MARKS]

Practice Paper 3: Calculator
Finally, here's Practice Paper 3 — you'll be pleased to know that you can use your
calculator for this one too. Take a look at the formulas page on p.244 if you need a
reminder of which formulas you need to learn and which you'll be given in the exam.

Candidate Surname		Candidate Forename(s)	
Centre Number	Candidate Number	Candidate Signature	

GCSE

Mathematics   Foundation Tier

Paper 3 (Calculator)

Practice Paper
Time allowed: 1 hour 30 minutes

You must have:
Pen, pencil, eraser, ruler, protractor, pair of compasses.
You may use tracing paper.

You **may use** a calculator.

Instructions to candidates
- Use **black** ink to write your answers.
- Write your name and other details in the spaces provided above.
- Answer **all** questions in the spaces provided.
- In calculations, show clearly how you worked out your answers.
- Do all rough work on the paper.
- Unless a question tells you otherwise, take the value of π to be 3.142,
  or use the π button on your calculator.

Information for candidates
- The marks available are given in brackets at the end of each question.
- You may get marks for method, even if your answer is incorrect.
- There are 28 questions in this paper. There are no blank pages.
- There are 80 marks available for this paper.

Answer ALL the questions.

Write your answers in the spaces provided.

You must show all of your working.

1   Write one of the signs <, =, or > on each answer line to make a true statement.

0.4   ....   0.34

$\frac{3}{4}$ 0.75

7% 0.7

[Total 2 marks]

2   The diagram shows part of a number line.

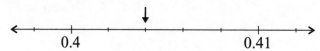

Circle the number the arrow points to.

0.42   0.44   0.402   0.404

[Total 1 mark]

3   Round 20 758 to the nearest 100.

....

[Total 1 mark]

4   What number is 12 less than –4.2?

....

[Total 1 mark]

5   Circle the number below that has exactly four factors.

2   3   5   8   9   12

[Total 1 mark]

6   24 pupils take a violin exam. Their marks are shown below.

110	103	115	134	121	98
128	112	107	112	125	132
114	102	125	93	120	120
106	111	99	98	127	115

The certificate each pupil receives depends upon their mark.

Result of exam	Mark
Fail	Under 100
Pass	100 – 119
Merit	120 – 129
Distinction	130 and above

(a) Complete the table to show the number of pupils achieving each result.
     The first row has been filled in for you.

Result of exam	Tally	Frequency
Fail	||||	4
Pass		
Merit		
Distinction		
	Total:	24

[2]

(b) What fraction of the pupils failed the exam? Give your fraction in its simplest form.

....

[2]

(c) Draw on the grid a suitable diagram to show the number of pupils achieving each result.

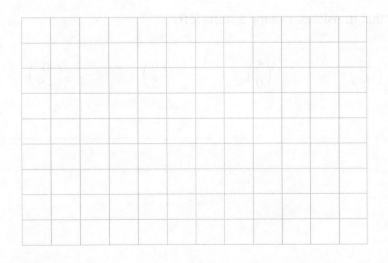

[3]

[Total 7 marks]

2

7   A pencil case contains 10 coloured pencils.

> 1 pencil is yellow.
> 2 pencils are red.
> The other pencils are either green or blue.

Carla picks one coloured pencil at random.
She has the same chance of picking a green pencil as a red pencil.

Circle the word that describes the probability of picking:

(a)   a black pencil,

> impossible   unlikely   evens   likely

[1]

(b)   a blue pencil.

> impossible   unlikely   evens   likely

[1]

[Total 2 marks]

8   Here are the names of four types of quadrilateral.

| Parallelogram | Square | Trapezium | Kite |

Choose from this list the quadrilateral that has:

(a)   exactly one pair of parallel sides,

....
[1]

(b)   no lines of symmetry, but rotational symmetry of order 2.

....
[1]

[Total 2 marks]

9   Circle the vector that translates a shape 5 units **left**.

$$\begin{pmatrix} -5 \\ 0 \end{pmatrix} \qquad \begin{pmatrix} 5 \\ 0 \end{pmatrix} \qquad \begin{pmatrix} 0 \\ 5 \end{pmatrix} \qquad \begin{pmatrix} 0 \\ -5 \end{pmatrix}$$

[Total 1 mark]

10 Rick multiplies three different numbers together and gets 90.
One of his numbers is a square number, and the other two are prime numbers.
What are the three numbers he uses?

....

[Total 3 marks]

11 Nigel sees this recipe for cupcakes.

> **Recipe for 12 cupcakes**
> 140 grams butter
> 140 grams flour
> 132 grams sugar
> 2 eggs
> 1 tablespoon milk

Nigel wants to make 30 cupcakes. How much sugar does he need?

.... g
[Total 2 marks]

12 Ajay buys some packets of ginger biscuits.
Jane buys some packets of shortbread biscuits.

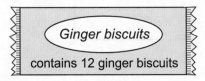 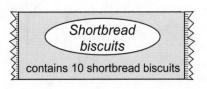

Ajay and Jane buy the same number of biscuits.

What is the smallest number of packets of shortbread biscuits Jane could have bought?

.... packets
[Total 3 marks]

13   Mary is preparing cream teas for 30 people.

Each person needs 2 scones, 1 tub of clotted cream and 1 small pot of jam.

She has £35 to buy everything.

     A pack of 10 scones costs £1.35
     A pack of 6 tubs of clotted cream costs £2.95
     Each small pot of jam costs 40p

Will she have enough money? Show how you work out your answer.

....

[Total 5 marks]

14   The grid shows part of two shapes, *A* and *B*.

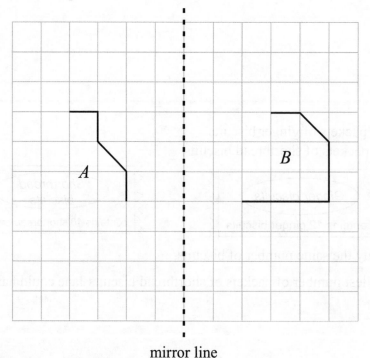

mirror line

B is the reflection of *A* in the mirror line.

Complete both shapes.

[Total 2 marks]

15 The diagram shows an object made from 8 centimetre cubes.

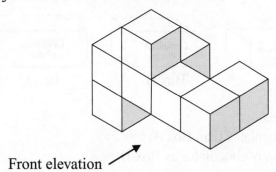

Front elevation

Draw on the grids below the plan view and the front elevation of the object.

Plan view

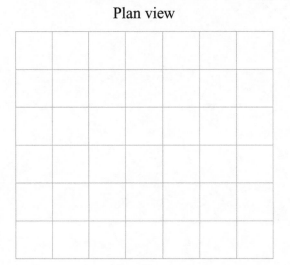

Front elevation

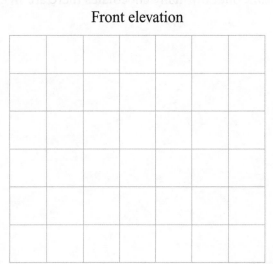

[Total 2 marks]

16 Two congruent trapeziums and two triangles fit inside a square of side 12 cm as shown.

12 cm

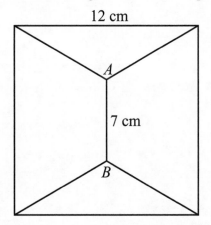

A

7 cm

B

Not to scale

AB = 7 cm

Work out the area of each trapezium.

.... cm²

[Total 2 marks]

Practice Paper 3

17 A chocolate manufacturer makes boxes of chocolates in three different sizes.

Box A   Box B   Box C

Box A contains c chocolates.
Box B contains 4 more chocolates than Box A.
Box C contains twice as many chocolates as Box B.
Altogether there are 60 chocolates.

Work out how many chocolates there are in each box.

Box A: ....

Box B: ....

Box C: ....

[Total 5 marks]

18 Simplify

(a) $y \times y \times y$

....
[1]

(b) $n^6 \div n^2$

....
[1]

(c) $(a^4)^3$

....
[1]

[Total 3 marks]

---

19 The diagram shows a ramp placed against two steps.

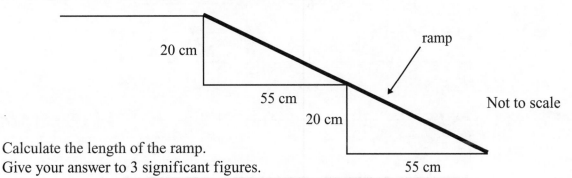

20 cm

55 cm

20 cm

55 cm

ramp

Not to scale

Calculate the length of the ramp.
Give your answer to 3 significant figures.

.... cm

[Total 3 marks]

---

20 A route between Guilford and Bath has a distance of 180 kilometres.
Dave drives from Guilford to Bath. He takes 3 hours.

Olivia drives the same route. Her average speed is 15 kilometres per hour faster than Dave's.

(a) How long does it take Olivia to drive from Guilford to Bath?
Give your answer in hours and minutes

.... hours .... minutes
[3]

(b) Why is it important to your calculation that Olivia drives the same route as Dave?

....

....
[1]

[Total 4 marks]

8

21 (a) Complete the table of values for $y = x^2 + x - 2$.

x	−3	−2	−1	0	1	2	3
y		0	−2	−2			10

[2]

(b) Draw on the grid the graph of $y = x^2 + x - 2$ for values of x between −3 and 3.

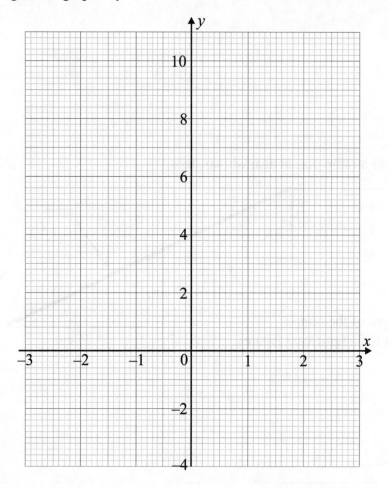

[2]

[Total 4 marks]

22 The ratio of angles in a triangle is $2:3:5$.
Show that this a right-angled triangle.

[Total 3 marks]

23 The values of four houses at the start of 2017 are shown.

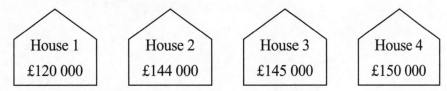

House 1 £120 000   House 2 £144 000   House 3 £145 000   House 4 £150 000

(a) Which house has a value 25% higher than House 1?

House ....
[1]

(b) At the start of 2019, the value of House 2 is £161 280.
Find the percentage increase in the value of House 2.

.... %
[3]

[Total 4 marks]

24 Anna and Carl each think of a sequence of numbers.

Anna's sequence

4th term = 17

Term-to-term rule is
Add 3

Carl's sequence

Term-to-term rule is
Add 6

The 1st term of Anna's sequence is double the 1st term of Carl's sequence.

Work out the 5th term of Carl's sequence.

....

[Total 3 marks]

25 (a) Factorise $x^2 + 7x - 18$.

....
[2]

(b) Solve the equation $x^2 + 7x - 18 = 0$.

$x = $ .... or $x = $ ....
[1]

[Total 3 marks]

26 George has two fair spinners.

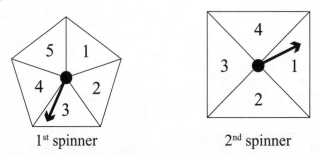

1st spinner   2nd spinner

He spins each spinner once and records whether the score is an odd or an even number.

(a) Complete the tree diagram to show the probabilities.

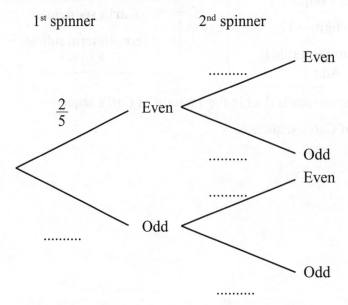

(b) Work out the probability that George spins two odd numbers.

....
[2]

[Total 4 marks]

11

27 The line *L* passes through the points (–2, –7) and (3, 8).
Find the equation of line *L*.

....
[Total 4 marks]

28 The grouped frequency table below shows the weights of 25 rabbits in a pet shop.

Weight (*w* g)	Frequency
$800 \leq w < 1000$	5
$1000 \leq w < 1200$	8
$1200 \leq w < 1400$	9
$1400 \leq w < 1600$	3

Estimate the mean weight.

.... g
[Total 3 marks]

[TOTAL FOR PAPER = 80 MARKS]

Answers

Answers to the <u>Warm-Up Questions</u>, <u>Exam Questions</u> and <u>Revision Questions</u> (which appear on the <u>green</u> and <u>yellow</u> pages of the book) and answers to the <u>Practice Papers</u> are given on the pages that follow.

Answers to the <u>Exam Practice Questions</u> that appear at the bottom of most other pages can be found <u>online</u>. They're included as a printable PDF with your free Online Edition — you'll find more info about how to get hold of this at the front of this book.

Section One — Number

Page 7 (Warm-up Questions)

1   20

2   £14.70

3   a) 957   b) 48

4   a) 0.245   b) 5

5   a) 387   b) 544   c) 121.8   d) 8.45

6   a) 13   b) 121   c) 62   d) 30

7   a) 12   b) –6   c) 9   d) –3

Page 8 (Exam Questions)

3   a) 81 *[1 mark]*

    b) 64 *[1 mark]*

4   E.g. $288 \div -3 = -96$
    $-96 \div 12 = -8$
    So the third number is –8
    [3 marks available — 1 mark for a correct method, 1 mark for at least one correct calculation, 1 mark for the correct answer]
    You could have worked out –3 × 12 (= –36) and divided by this instead, but the division would be pretty tricky.

5   £200 – £5 = £195
    $15 \overline{)1^{1}9^{4}5}$, so each ticket costs £13
    with 0 1 3 above
    [3 marks available — 1 mark for subtracting £5 from £200, 1 mark for dividing £195 by 15, 1 mark for the correct final answer]

6   Total miles travelled = (30 × 2) + (28 × 2) + (39 × 2) + (40 × 2)
                          = 60 + 56 + 78 + 80
                          = 274 miles
    Expenses for miles travelled = 274 × 30p = 8220p = £82.20
    Expenses for food = 4 × £8 = £32
    Total expenses = £82.20 + £32 = £114.20
    [4 marks available — 1 mark for finding total miles, 1 mark for multiplying total miles by 30 or 0.3(0), 1 mark for finding food expenses, 1 mark for the correct final answer]

Page 13 (Warm-up Questions)

1   31, 37

2   27 ÷ 3 = 9. So 27 is not a prime number because it divides by 3 and 9.

3   1, 2, 4, 5, 8, 10, 20, 40

4   2 and 5

5   20 *(multiples of 4 are: 4, 8, 12, 16, 20, ...,*
       multiples of 5 are: 5, 10, 15, 20, ...)

6   8 *(factors of 32 are: 1, 2, 4, 8, 16, 32,*
      factors of 88 are: 1, 2, 4, 8, 11, 22, 44, 88)

Page 14 (Exam Questions)

3   Jasmine is incorrect as there are four prime numbers (101, 103, 107 and 109) between 100 and 110.
    [2 marks available — 1 mark for stating that Jasmine is incorrect, 1 mark for providing evidence]
    Writing one prime number between 100 and 110 is enough.

4   E.g. 37 (3 + 7 = 10, which is 1 more than 9, a square number)
    [2 marks available — 2 marks for a correct answer, otherwise 1 mark for a prime of two or more digits]

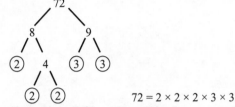

5
$$72 = 2 \times 2 \times 2 \times 3 \times 3$$
    [2 marks available — 1 mark for a correct method, 1 mark for all prime factors correct]

6   a) LCM = $3^7 \times 7^3 \times 11^2$ *[1 mark]*

    b) HCF = $3^4 \times 11$ *[1 mark]*

Page 20 (Warm-up Questions)

1   $\frac{3}{4}$

2   $\frac{2}{6}$ and $\frac{5}{15}$

3   a) $\frac{4}{15}$

    b) $\frac{2}{5} \div \frac{2}{3} = \frac{2}{5} \times \frac{3}{2} = \frac{6}{10} = \frac{3}{5}$

    c) $\frac{2}{5} + \frac{2}{3} = \frac{6}{15} + \frac{10}{15} = \frac{16}{15} = 1\frac{1}{15}$

    d) $\frac{2}{3} - \frac{2}{5} = \frac{10}{15} - \frac{6}{15} = \frac{4}{15}$

4   0.7

5   $66.\dot{6}\%$ (66.666...%) or $66\frac{2}{3}\%$

6   $\frac{2}{5}$

7   $0.\dot{2}8571\dot{4}$

Page 21 (Exam Questions)

3   (12 400 ÷ 8) × 3 = 1550 × 3 = 4650
    [2 marks available — 1 mark for dividing by 8 or multiplying by 3, 1 mark for the correct answer]

4   65% = 0.65, $\frac{2}{3} = 0.666...$, $\frac{33}{50} = 0.66$
    So order is 0.065, 65%, $\frac{33}{50}, \frac{2}{3}$
    [2 marks available — 2 marks for all four numbers in the correct order, otherwise 1 mark for writing the numbers in the same form (either decimals, percentages or fractions)]

5   a) $1\frac{1}{8} \times 2\frac{2}{5} = \frac{9}{8} \times \frac{12}{5}$ *[1 mark]* $= \frac{108}{40}$ *[1 mark]*
        $= 2\frac{7}{10}$ *[1 mark]*
        [3 marks available in total — as above]

    b) $1\frac{3}{4} \div \frac{7}{9} = \frac{7}{4} \times \frac{9}{7}$ *[1 mark]* $= \frac{63}{28}$ or $\frac{9}{4}$ *[1 mark]*
        $= 2\frac{1}{4}$ *[1 mark]*
        [3 marks available in total — as above]

6   $\frac{1}{4}$ = 25%, so Jenny pays 1 – 25% – 20% – 20%
                                      = 1 – 65% = 35% *[1 mark]*
    £17.50 = 35% *[1 mark]*, so 1% = £17.50 ÷ 35 = £0.50.
    The total bill was £0.50 × 100 *[1 mark]* = £50 *[1 mark]*.
    [4 marks available in total — as above]

Page 26 (Warm-up Questions)

1   a) 3.2   b) 1.8
    c) 2.3   d) 0.5
    e) 9.8

2   a) 3   b) 5
    c) 2   d) 7
    e) 3

3 a) 350 *(the decider is 2, so keep the 5, and fill the missing place with zero)*

b) 500 *(the decider is 6, so round the 4 up to 5, and fill the missing places with zeros)*

c) 12.4 *(the decider is 8, so round the 3 up to 4)*

d) 0.036 *(the decider is 6, so round the 5 up to 6)*

4 a) 2900 b) 500

c) 100

5 a) 100 *(This is approximately (30 – 10) × 5)*

b) Overestimate *(you rounded all the numbers up, so your estimate will be bigger than the actual answer)*

6 567.5 ml

7 a) $0.9145 \leq x < 0.9155$

b) $1115 \leq y < 1125$

8 a) 37.9

b) 2.1

Page 27 (Exam Questions)

2 a) 428.6 light years *[1 mark]*

b) 430 light years *[1 mark]*

3 a) E.g. $(£4.95 × 28) + (£11 × 19) \approx (£5 × 30) + (£10 × 20)$
$= £150 + £200 = £350$
[2 marks available — 1 mark for rounding each value sensibly, 1 mark for a sensible estimate]

b) E.g. This is a sensible estimate as it is very close to the actual value of £347.60 *[1 mark]*.

4 E.g. $\dfrac{12.2 \times 1.86}{0.19} \approx \dfrac{10 \times 2}{0.2} = \dfrac{20}{0.2} = 100$
[2 marks available — 1 mark for rounding to suitable values, 1 mark for the correct final answer using your values]

5 Minimum weight = 56.5 kg *[1 mark]*
Maximum weight = 57.5 kg *[1 mark]*
[2 marks available in total — as above]

6 Smallest possible value of $a = 3.8 – 0.05 = 3.75$
Largest possible value of $a = 3.8 + 0.05 = 3.85$
So error interval is $3.75 \leq a < 3.85$
[2 marks available — 1 mark for $3.75 \leq a$, 1 mark for $a < 3.85$]

Page 32 (Warm-up Questions)

1 a) 0.4096

b) 49

c) 10^6

2 a) $4^6 (= 4096)$ b) $5^6 (= 15\ 625)$

c) x^6

3 a) 49 b) $\dfrac{1}{27}$

4 a) 15

b) 19

c) 6

5 10 cm

6 27.18411

7 a) $2.89 × 10^6$

b) 0.0000711

8 $2.1 × 10^4$

9 a) $1.2 × 10^2$

b) $1.43 × 10^{-6}$

Page 33 (Exam Questions)

2 $\sqrt{6.25} = 2.5$ cm *[1 mark]*

3 a) $A = 4.834 × 10^9 = 4\ 834\ 000\ 000$ *[1 mark]*

b) C, B, A $(5.21 × 10^3, 2.4 × 10^5, 4.834 × 10^9)$ *[1 mark]*

4 $\dfrac{3^4 \times 3^7}{3^6} = \dfrac{3^{(4+7)}}{3^6} = \dfrac{3^{11}}{3^6} = 3^{(11-6)} = 3^5$
[2 marks available — 1 mark for a correct attempt at adding or subtracting powers, 1 mark for the correct final answer]

5 a) $6^{(5-3)} = 6^2 = 36$ *[1 mark]*

b) $(2^4 × 2^7) = 2^{(4+7)} = 2^{11}$
$(2^3 × 2^2) = 2^{(3+2)} = 2^5$, so $(2^3 × 2^2)^2 = (2^5)^2 = 2^{10}$
So $(2^4 × 2^7) ÷ (2^3 × 2^2)^2 = 2^{11} ÷ 2^{10} = 2^1 = 2$
[2 marks available — 1 mark if each bracket has been correctly simplified, 1 mark for the correct answer]

6 time (s) = distance (miles) ÷ speed (miles/s)
$= (9 × 10^7) ÷ (2 × 10^5)$ seconds *[1 mark]*
$= 450$ seconds *[1 mark]*
[2 marks available in total — as above]

Page 34 (Revision Questions)

1 A square number is a whole number multiplied by itself. The first ten are: 1, 4, 9, 16, 25, 36, 49, 64, 81 and 100.

2 0.1

3 £38

4 a) £120 b) £0.50 = 50p

5 a) 1377 b) 26

c) 62.7 d) 0.35

6 a) –16 b) 7 c) 20

7 41, 43, 47, 53, 59

8 The multiples of a number are its times table.

a) 10, 20, 30, 40, 50, 60

b) 4, 8, 12, 16, 20, 24

9 a) $210 = 2 × 3 × 5 × 7$

b) $1050 = 2 × 3 × 5 × 5 × 7$
$= 2 × 3 × 5^2 × 7$

10 a) 14 b) 40

11 a) $\dfrac{25}{16}$ or $1\dfrac{9}{16}$ b) $\dfrac{44}{15}$ or $2\dfrac{14}{15}$

c) $\dfrac{23}{8}$ or $2\dfrac{7}{8}$ d) $\dfrac{11}{21}$

12 a) 320 b) £60

13 Amy

14 a) i) $\dfrac{4}{100} = \dfrac{1}{25}$ ii) 4%

b) i) $\dfrac{65}{100} = \dfrac{13}{20}$ ii) 0.65

15 a) Recurring decimals have a pattern of numbers which repeats forever.

b) $0.\dot{2}$

16 a) 17.7

b) 6700

c) 4 000 000

17 a) 100

b) 1400

18 $150 \leq x < 250$

19 7^5

20 $\dfrac{1}{25}$

21 a) 11 b) 4

c) 56 d) 10^5

22 a) 421.875 b) 4.8

c) 8 d) 11

23 1. The front number must always be between 1 and 10.

2. The power of 10, n, is how far the decimal point moves.

3. n is positive for big numbers, and negative for small numbers.

24 a) $3.56 × 10^9$ b) 0.00000275

25 a) $2 × 10^3$ b) $1.2 × 10^{12}$

Section Two — Algebra

Page 39 (Warm-up Questions)

1 a) $-2a$ b) $7p - 10q$ c) $6 + 7\sqrt{7}$

2 a) $-24mn$ b) $-6r + 33$ c) $5x^2 - 35xy$

3 $3(x + 4) - 6(3 - 2x) = 3x + 12 - 18 + 12x = 15x - 6$

4 a) $x^2 - 3x - 18$ b) $5a^2 - 18a - 8$
   c) $x^2 + 16x + 64$ d) $4y^2 - 12y + 9$

5 a) $3(x - 3)$ b) $4y(1 + 5y)$
   c) $(x + 4)(x - 4)$ d) $(4x + 5y)(4x - 5y)$

Page 40 (Exam Questions)

3 a) $4p$ *[1 mark]*
   b) $2m$ *[1 mark]*
   c) $4p + 3r$
      [2 marks available — 1 mark for 4p and 1 mark for 3r]

4 $6x + 3 = (3 \times 2x) + (3 \times 1) = 3(2x + 1)$ *[1 mark]*

5 a) $(x + 2)(x + 4) = x^2 + 4x + 2x + 8 = x^2 + 6x + 8$
      [2 marks available — 1 mark for expanding the brackets correctly, 1 mark for simplifying]
   b) $(y + 3)(y - 3) = y^2 - 3y + 3y - 9 = y^2 - 9$
      [2 marks available — 1 mark for expanding the brackets correctly, 1 mark for simplifying]
   c) $(2z - 1)(z - 5) = 2z^2 - 10z - z + 5 = 2z^2 - 11z + 5$
      [2 marks available — 1 mark for expanding the brackets correctly, 1 mark for simplifying]

6 a) $x^2 - 49 = x^2 - 7^2 = (x + 7)(x - 7)$
      [2 marks available — 2 marks for the correct final answer, otherwise 1 mark for attempting to use the difference of two squares]
   b) $9x^2 - 100 = (3x)^2 - 10^2 = (3x + 10)(3x - 10)$
      [2 marks available — 2 marks for the correct final answer, otherwise 1 mark for attempting to use the difference of two squares]
   c) $y^2 - m^2 = (y + m)(y - m)$
      [2 marks available — 2 marks for the correct final answer, otherwise 1 mark for attempting to use the difference of two squares]

Page 47 (Warm-up Questions)

1 a) $x = 2$ b) $x = 11$ c) $x = 6$ d) $x = 20$

2 a) $x = 4$ b) $y = 11$

3 $s = 12$

4 a) $y = 25$ b) $x = 11$

5 Total sweets in jar $= 7b + 10$ (where b = number of blue sweets)

6 20

7 $x = 3$

8 $b = 5a - 4$

9 $q = \dfrac{p + 8}{3}$

Page 48 (Exam Questions)

2 a) $S = 4m^2 + 2.5n$
      $S = (4 \times 2 \times 2) + (2.5 \times 10)$
      $S = 16 + 25 = 41$
      [2 marks available — 1 mark for correct substitution of m and n, 1 mark for correct final answer]
   b) $S = 4m^2 + 2.5n$
      $S = (4 \times 6.5 \times 6.5) + (2.5 \times 4)$
      $S = 169 + 10 = 179$
      [2 marks available — 1 mark for correct substitution of m and n, 1 mark for correct final answer]

3 a) $40 - 3x = 17x$
      $40 = 20x$ *[1 mark]*
      $x = 40 \div 20 = 2$ *[1 mark]*
      [2 marks available in total — as above]

b) $2y - 5 = 3y - 12$
   $-5 + 12 = 3y - 2y$ *[1 mark]*
   $y = 7$ *[1 mark]*
   [2 marks available in total — as above]

4 The sides of an equilateral triangle are all the same length, so
   $4(x - 1) = 3x + 5$ *[1 mark]*
   $4x - 4 = 3x + 5$
   $x = 9$ *[1 mark]*
   So each side is $(3 \times 9) + 5 = 32$ cm long *[1 mark]*.
   [3 marks available in total — as above]
   To check your answer, put your value of x into the expression for the other side of the triangle — you should get the same answer.

5 $\dfrac{a + 2}{3} = b - 1$
   $a + 2 = 3b - 3$ *[1 mark]*
   $a = 3b - 5$ *[1 mark]*
   [2 marks available in total — as above]

6 Call the number of Whitewater fans f. Redwood fans $= 3 \times f = 3f$.
   Difference $= 3f - f = 2f$, so $2f = 7000$, so $f = 3500$.
   Total fans $= 3f + f = 4f = 4 \times 3500 = 14\,000$.
   [3 marks available — 1 mark for the expressions for the number of fans for each team, 1 mark for forming and solving the equation to find f, 1 mark for the correct answer]

Page 52 (Warm-up Questions)

1 a) 18, 22   b) 81, 243   c) 17, 23

2 a) Rule = square the number of the term
   b) 36, 49
   c) Yes, because $100 = 10 \times 10 = 10^2$ so it is a square number.

3 a) Arithmetic — to get from one term to the next, you add the same number to the previous term each time.
   b) Geometric — to get from one term to the next, you multiply the previous term by the same number each time.

4 a) $-4n + 31$ b) -49

5 $n = -2, -1, 0, 1, 2, 3, 4$

6 a) $x < 5$ b) $x \geq 7$

Page 53 (Exam Questions)

2 $-3, -2, -1, 0, 1$
   [2 marks available — 2 marks for all 5 numbers correct, otherwise 1 mark for the correct answer with one number missing or one number incorrect]

3 a)
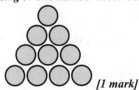
   [1 mark]
   b) The number of circles added increases by one each time, so the tenth triangle number is:
      $1 + 2 + 3 + 4 + 5 + 6 + 7 + 8 + 9 + 10 = 55$.
      [2 marks available — 1 mark for 55 and 1 mark for correct reasoning]

4 Second term $= 7 - 3 = 4$
   Fourth term $= 4 + 7 = 11$
   Fifth term $= 7 + 11 = 18$
   [2 marks available — 2 marks for all three terms correct, otherwise 1 mark for at least one term correct]

5 2   6   12   20
     +4 +6   +8

   The difference is increasing by 2, so the next term is:
   $20 + 10 = 30$
   [2 marks available — 1 mark for spotting the pattern, 1 mark for the correct answer]

6 Largest possible value of $p = 45$
   Smallest possible value of $q = 26$ *[1 mark for both]*
   Largest possible value of $p - q = 45 - 26 = 19$ *[1 mark]*.
   [2 marks available in total — as above]

Page 57 (Warm-up Questions)

1. a) $(x-4)(x-2)$ b) $(x+1)(x-3)$
   c) $(x+4)(x+3)$
2. a) $x=-2$ or $x=-5$ b) $x=2$ or $x=-7$
   c) $x=2$ or $x=3$
3. $x=1, y=4$
4. $x=-1, y=3$
5. a) E.g. $2^2 \times 5^2 = 4 \times 25 = 100 (= 10^2)$
   b) E.g. $2+7=9 (9 = 3 \times 3)$
6. LHS: $(x+2)^2 + (x-2)^2 = (x^2 + 4x + 4) + (x^2 - 4x + 4)$
   $= 2x^2 + 8$
   $= 2(x^2 + 4) = $ RHS

Page 58 (Exam Questions)

3. a) 16 is a factor of 48 *[1 mark]*
   b) E.g. $4 + 16 = 20$, which is even *[1 mark]*
   c) E.g. 38 is not a multiple of 4, 6 or 8 *[1 mark]*

4. $x + 3y = 11$ (1) $\xrightarrow{\times 3}$ $3x + 9y = 33$ (3) *[1 mark]*
   $3x + y = 9$ (2)
   (3) − (2):
   $3x + 9y = 33$ $x + 3y = 11$
   $- \underline{3x + y = 9}$ $x + (3 \times 3) = 11$
       $8y = 24$ $x = 11 - 9$
       $y = 3$ *[1 mark]* $x = 2$ *[1 mark]*
   [3 marks available in total — as above]

5. 6 and 2 multiply to give 12 and subtract to give 4,
   so if $x^2 + 4x - 12 = 0$,
   $(x + 6)(x - 2) = 0$
   [1 mark for correct numbers in brackets,
   1 mark for correct signs]
   $x + 6 = 0$ or $x - 2 = 0$
   $x = -6$ or $x = 2$
   [1 mark for both solutions]
   [3 marks available in total — as above]

6. $2(18 + 3q) + 3(3 + q) = 36 + 6q + 9 + 3q$
   $= 9q + 45$
   $= 9(q + 5)$

   $2(18 + 3q) + 3(3 + q)$ can be written as $9 \times$ a whole number
   (where the whole number is $(q + 5)$), so it is a multiple of 9.
   [3 marks available — 1 mark for expanding brackets and
   simplifying, 1 mark for writing the expression as 9(q + 5),
   1 mark for explaining why this is a multiple of 9]

Page 59 (Revision Questions)

1. a) $3e$ b) $12f$
2. a) $7x - y$ b) $3a + 9$
3. a) m^3 b) $7pq$ c) $18xy$
4. a) $6x + 18$ b) $-9x + 12$ c) $5x - x^2$
5. $6x$
6. a) $2x^2 - x - 10$ b) $25y^2 + 20y + 4$
7. Putting in brackets (the opposite of multiplying out brackets).
8. a) $8(x + 3)$ b) $9x(2x + 3)$ c) $(6x + 9y)(6x - 9y)$
9. a) $x = 7$ b) $x = 16$ c) $x = 3$
10. a) $x = 4$ b) $x = 2$ c) $x = 3$
11. $Q = 8$
12. 14
13. 37 marbles
14. $6x$ cm
15. $v = \dfrac{W - 5}{4}$
16. a) 31, rule is add 7
    b) 256, rule is multiply by 4
    c) 19, rule is add previous two terms.
17. $6n - 2$
18. Yes, it's the 5th term.

19. a) x is greater than minus seven.
    b) x is less than or equal to six.
20. $k = 1, 2, 3, 4, 5, 6, 7$
21. a) $x < 10$ b) $x \leq 7$
22. a) $(x + 2)(x + 8)$ b) $(x + 1)(x - 7)$
23. $x = -6$ or $x = 3$
24. $x = 2, y = 5$
25. E.g. 19
26. $3(y + 2) + 2(y + 6)$
    $= 3y + 6 + 2y + 12$
    $= 5y + 18 = 5(y + 3) + 3.$
    $5(y + 3)$ is a multiple of 5, so $5(y + 3) + 3$ is not a multiple of 5.

Section Three — Graphs

Page 66 (Warm-up Questions)

1. a)
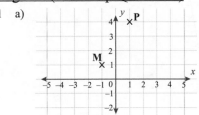
   b) Midpoint = $(0, 2.5)$
2. a) Yes   b) No   c) Yes   d) Yes
   Try to rearrange each equation into the form $y = mx + c$ —
   if you can do this then it's a straight line, if you can't it's not.

3.
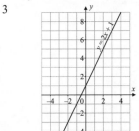

4.
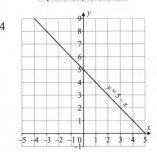

5. $\dfrac{15}{16}$
6. 3
7. a) $y = -2x - 3$ b) $y = -2x + 9$

Page 67 (Exam Questions)

2. a) $\left(\dfrac{1 + 3}{2}, \dfrac{3 + (-1)}{2}\right) = (2, 1)$
   [2 marks available — 1 mark for correct method and
   1 mark for correct final answer]
   A correct method here is to find the averages of the x- and
   y-coordinates. Or, you could identify the midpoint of AB on the
   graph to get your answer — but the first way is much safer.
   b) Comparing coordinates of point **C** and midpoint of **CD**:
   x-distance = $2 - 0 = 2$
   y-distance = $1 - -1 = 2$
   So to get from the midpoint to point **D**, move up 2 and right 2.
   So point **D** is $(2 + 2, 1 + 2) = (4, 3)$
   [2 marks available — 1 mark for each correct coordinate]

3　Find the gradient: $\dfrac{\text{change in } y}{\text{change in } x} = \dfrac{4-1}{1-0} = 3$

Line crosses y-axis at 1, so equation of line is $y = 3x + 1$.
[3 marks available — 3 marks for a fully correct answer, otherwise 2 marks for a correct gradient, or 1 mark for a correct method to find the gradient]

4　The lines are parallel, so their gradients are equal: m = 4 *[1 mark]*
When $x = -1$, $y = 0$, so put this into $y = 4x + c$
$0 = (4 \times -1) + c$, so c = 4 *[1 mark]*
So equation of line is $y = 4x + 4$ *[1 mark]*
[3 marks available in total — as above]

Page 74 (Warm-up Questions)

1　a)
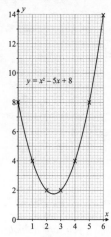

b)　(2.5, 1.75)

2
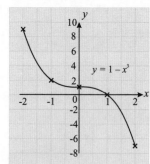

3　$x = 3$, $y = 1$

4　a)　36 litres (allow 35.5-36.5 litres)
　　b)　4.4 gallons (allow 4.3-4.5 gallons)

5　a)　10 seconds　　　b)　4 m/s

6　Graph B

Pages 75-76 (Exam Questions)

2　a)
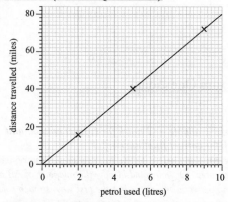

[2 marks available — 1 mark for all points plotted correctly, 1 mark for straight line joining points]

　　b)　Gradient = $\dfrac{\text{change in } y}{\text{change in } x} = \dfrac{80-0}{10-0} = 8$

[2 marks available — 1 mark for correct method to find gradient, 1 mark for correct answer]

c)　Distance travelled in miles per litre of petrol used *[1 mark]*

3　a)　0 *[1 mark]*
　　b)　−2, 0 *[1 mark]*
　　c)　(−1, −1) *[1 mark]*

4　a)
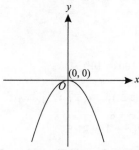

[2 marks available — 1 mark for correct shape, 1 mark for labelling (0, 0)]

　　b)
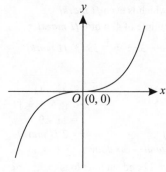

[2 marks available — 1 mark for correct shape, 1 mark for labelling (0, 0)]

5　a)　$x = 1$ *[1 mark]*
　　b)
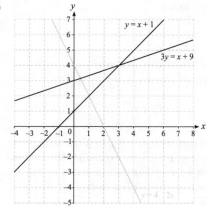

$x = 3$, $y = 4$
[3 marks available — 2 marks for correctly drawing the line $3y = x + 9$, 1 mark for the correct answer]

Page 77 (Revision Questions)

1　A(5, −3),　B(4, 0),　C(0, 3), D(−4, 5), E(−2, −3)

2　Midpoint = (2, 1.5)

3
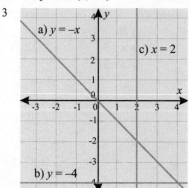

4 E.g.

x	−2	0	2
y	6	−2	−10

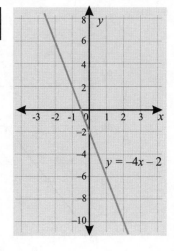

$y = -4x - 2$

5 a) 2   b) $y = x + 15$

6 $y = 2x - 6$

7 $y = x - 9$

8 They are both "bucket shaped" graphs. $y = x^2 - 8$ is like a "u" whereas $y = -x^2 + 2$ is like an "n" (or an upturned bucket).

9 a)

x	−3	−2	−1	0	1
y	0	−2	−2	0	4

b)

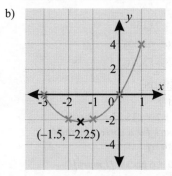

(−1.5, −2.25)

10 a) A graph with a "wiggle" in the middle. E.g.

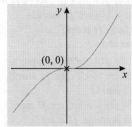

(0, 0)

b) A graph made up of two curves in opposite corners. The curves are symmetrical about the lines $y = x$ and $y = -x$. E.g.

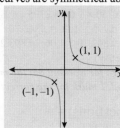

(1, 1)

(−1, −1)

11

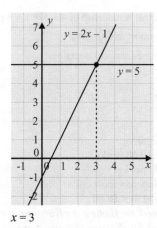

$y = 2x - 1$

$y = 5$

$x = 3$

12 The object has stopped.

13 a) Ben drove fastest on his way home.   b) 15 minutes

14 You would have to find the gradient, as the gradient = the rate of change.

15 a) 20 minutes   b) £20

    c) 26 minutes   d) 67p — allow between 65p and 69p

Section Four — Ratio, Proportion, and Rates of Change

Page 81 (Warm-up Questions)

1 a) 1 : 2   b) 4 : 9   c) 2 : 9   d) 16 : 7   e) 5 : 4

2 1 : 4.4

3 300 g ÷ 2 = 150 g; 150 g × 3 = 450 g of flour

4 a) 12 : 6 = 2 : 1   b) $\frac{12}{18} = \frac{2}{3}$

5 20

6 £1000 : £1400

7 45, 60, 75
   (3 + 4 + 5 = 12 parts, so 180 ÷ 12 = 15 per part.)

Page 82 (Exam Questions)

2 a) Shortest side of shape A = 3 units
       Shortest side of shape B = 6 units
       Ratio of shortest sides = 3 : 6 = 1 : 2
       [2 marks available — 1 mark for finding the shortest sides of the triangles, 1 mark for the correct answer]

   b) Area of shape A = $\frac{1}{2}$ × 3 × 4 = 6 square units *[1 mark]*

       Area of shape B = $\frac{1}{2}$ × 6 × 8 = 24 square units *[1 mark]*

       Ratio of areas = 6 : 24 = 1 : 4 *[1 mark]*
       [3 marks available in total — as above]

3 Donations account for 14 parts = £21 000
   So 1 part = £21 000 ÷ 14 = £1500 *[1 mark]*
   Bills are 5 parts so cost £1500 × 5 = £7500 *[1 mark]*
   £21 000 − £7500 = £13 500 *[1 mark]*
   [3 marks available in total — as above]
   Careful here — you are given a part : whole ratio in the question.

4   Mr Appleseed's Supercompost is made up of $4 + 3 + 1 = 8$ parts, so contains: $\frac{4}{8}$ soil, $\frac{3}{8}$ compost and $\frac{1}{8}$ grit.

16 kg of Mr Appleseed's Supercompost contains:

$\frac{4}{8} \times 16 = 8$ kg of soil

$\frac{3}{8} \times 16 = 6$ kg of compost

$\frac{1}{8} \times 16 = 2$ kg of grit

Soil costs £8 ÷ 40 = £0.20 per kg.
Compost costs £15 ÷ 25 = £0.60 per kg.
Grit costs £12 ÷ 15 = £0.80 per kg.
16 kg of Mr Appleseed's Supercompost costs:
$(8 \times 0.2) + (6 \times 0.6) + (2 \times 0.8) = £6.80$
[5 marks available — 1 mark for finding the fractions of each material in the mix, 1 mark for the correct mass of one material, 1 mark for the correct masses for the other two materials, 1 mark for working out the price per kg for each material, 1 mark for the correct answer]

Page 86 (Warm-up Questions)

1   £1.28

2   a) £12   b) 13

3   525 g
In the 250 g jar you get 250 g ÷ 125p = 2 g per p,
in the 350 g jar you get 350 g ÷ 210p = 1.666... g per p,
in the 525 g jar you get 525 g ÷ 250p = 2.1 g per p.

4   £210

5   Direct proportion graphs are straight lines and they go through the origin.

6
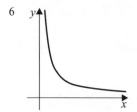

Page 87 (Exam Questions)

3   250 ml bottle: 250 ÷ 200 = 1.25 ml per penny
330 ml bottle: 330 ÷ 275 = 1.2 ml per penny
525 ml bottle: 525 ÷ 375 = 1.4 ml per penny
So the 525 ml bottle is the best value for money.
[3 marks available — 3 marks for finding the correct amounts per penny for all three bottles and the correct answer, otherwise 2 marks for two correct amounts per penny or 1 mark for one correct amount per penny]
You could also compare the cost per ml of each bottle.

4   1 bottle of water costs £52.50 ÷ 42 = £1.25 *[1 mark]*
There are £35 ÷ £1.25 = 28 girls in the club *[1 mark]*
[2 marks available in total — as above]

5   a) 250 people can be catered for 6 days
1 person can be catered for 6 × 250 = 1500 days
300 people can be catered for 1500 ÷ 300 = 5 days
[2 marks available — 1 mark for a correct method, 1 mark for the correct answer]

     b) For a 1-day cruise it could cater for 6 × 250 = 1500 people
For a 2-day cruise it could cater for 1500 ÷ 2 = 750 people
So it can cater for 750 – 250 = 500 more people
[3 marks available — 1 mark for a correct method to find the number of people catered for on a 2-day cruise, 1 mark for the correct number of people catered for on a 2-day cruise, 1 mark for the correct final answer]

6   a)
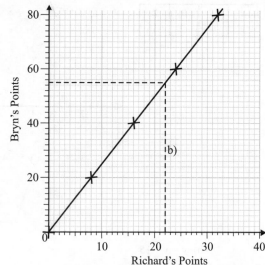

[2 marks available — 1 mark for two points marked correctly, 1 mark for the correct straight line]
Use the ratio to work out the coordinates of a few points to plot.
E.g. If Richard scored 8 points, Bryn scored $8 \times \frac{5}{2} = 20$ points.

     b) 55 points (see graph) *[1 mark]*

Page 93 (Warm-up Questions)

1   £17   2   74%

3   9   4   £138

5   £205   6   62.5%

7   £3   8   £3376.53

Pages 94 (Exam Questions)

2   20% increase = 1 + 0.2 = 1.2
20% increase of £33.25 = 1.2 × £33.25 = £39.90
[2 marks available — 1 mark for a correct method, 1 mark for the correct answer]

3   He normally gets 240 ÷ 40 = 6 packs *[1 mark]*
40% cheaper = 1 – 0.4 = 0.6
So the stickers are 40p × 0.6 = 24p per pack this week *[1 mark]*
He can buy 240 ÷ 24 = 10 packs this week *[1 mark]*
So he can get 10 – 6 = 4 more packs *[1 mark]*
[4 marks available in total — as above]

4   A ratio of 3 : 7 means 3 out of 10 = 30% of the animals are cats
40% of 30% = 0.4 × 30% = 12% are black cats *[1 mark]*
100% – 30% = 70% are dogs *[1 mark]*
50% of 70% = 0.5 × 70% = 35% are black dogs *[1 mark]*
So, 35% + 12% = 47% are black animals *[1 mark]*
[4 marks available in total — as above]

5   Multiplier = 1 + 0.06 = 1.06
After 1 year she will owe: £750 × 1.06 = £795
After 2 years she will owe: £795 × 1.06 = £842.70
After 3 years she will owe: £842.70 × 1.06 = £893.262
= £893.26 (to the nearest penny)
[3 marks available — 1 mark for working out the multiplier, 1 mark for a correct method, 1 mark for the correct answer]

Page 100 (Warm-up Questions)

1   a) 65 mm   b) 0.25 kg

2   160 kg

3   3 feet 10 inches

4   a) 320 km   b) 4 feet
To do these conversions, find the conversion factor, then multiply and divide by it. Then choose the most sensible answer.

5   a) 230 000 cm²   b) 3.45 m²

6   99 minutes

7   4.05 pm

8   0.16 m²

9   20 m/s

10 375 cm³

Pages 101 (Exam Questions)

2   4.30 pm till 5.00 pm is 30 minutes.
    5.00 pm till 7.00 pm is 2 hours.
    7.00 pm till 7.15 pm is 15 minutes.
    So they spend:
    2 hours + 30 minutes + 15 minutes = 2 hours 45 minutes
    2 hours 45 minutes = 2.75 hours
    $2.75 \times 12 = 33$ hours
    33 hours + 7 hours 10 minutes = 40 hours 10 minutes
    [4 marks available — 1 mark for a correct method to find the time from 4.30 pm till 7.15 pm, 1 mark for finding the correct time from 4.30 pm till 7.15 pm, 1 mark for the correct total time for the first 12 days, 1 mark for the correct answer]

3   64 pints = $64 \div 8 = 8$ gallons *[1 mark]*
    8 gallons = 4×2 gallons $\approx 4 \times 9$ litres *[1 mark]*
    = 36 litres *[1 mark]*
    [3 marks available in total — as above]

4   One book weighs 0.55 lb so 8 books will weigh
    8×0.55 lb = 4.4 lb *[1 mark]*
    1 kg $\approx$ 2.2 lb
    $4.4 \div 2.2 = 2$
    So the eight books weigh 2 kg. *[1 mark]*
    1 kg = 1000 g
    $2 \times 1000 = 2000$
    So the books weigh 2000 g. *[1 mark]*
    For 100 g postage is £0.50, so for 2000 g postage is
    £0.50 $\times$ 20 = £10. *[1 mark]*
    [4 marks available in total — as above]

5   Area of face A = 2 m $\times$ 4 m = 8 m² *[1 mark]*
    Pressure = Force $\div$ Area = 40 N $\div$ 8 m² *[1 mark]*
    = 5 N/m² *[1 mark]*
    [3 marks available in total — as above]

Page 102 (Revision Questions)

1   a) 9:11   b) 3.5:1

2   80 blue scarves

3   $\frac{7}{2}$ or 3.5

4   a) $\frac{5}{25}$ or $\frac{1}{5}$ b) 384

5   51 ml olive oil, 1020 g tomatoes, 25.5 g garlic powder,
    204 g onions

6   960 flowers

7   See p.84

8   The 500 ml tin

9   18

10 a) 19   b) 114   c) 21.05% (2 d.p.)   d) 475%

11 percentage change = (change $\div$ original) $\times$ 100

12 35% decrease

13 17.6 m

14 2%

15 a) £117.13 (to the nearest penny)   b) 6 years

16 a) 5600 cm³   b) 240 cm   c) 336 hours
    d) 12 000 000 cm³   e) 12.8 cm²   f) 2750 mm³

17 9.48 pm

18 67.2 km/h

19 12 500 cm³

20 11 m²

Section Five — Shapes and Area

Page 110 (Warm-up Questions)

1   Ɛ   1 line of symmetry, rotational symmetry order 1
    W   1 line of symmetry, rotational symmetry order 1
    ┼   2 lines of symmetry, rotational symmetry order 2
    Đ   1 line of symmetry, rotational symmetry order 1
    Q   0 lines of symmetry, rotational symmetry order 1

2   An equilateral triangle has 3 equal sides, 3 equal angles of 60°,
    3 lines of symmetry and rotational symmetry of order 3.

3   A kite has 1 line of symmetry.

4   a) **B** and **E** are similar.
    b) **A** and **D** are congruent.

5   A → B — rotation of 90° clockwise about the origin.
    B → C — reflection in the line $y = x$.
    C → A — reflection in the y-axis.
    A → D — translation by the vector $\begin{pmatrix} -9 \\ -7 \end{pmatrix}$.

6   (9, –3)

Page 111 (Exam Questions)

2   $x = 3$ *[1 mark]*

3
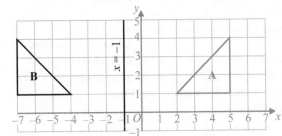

    [2 marks available — 2 marks for correct reflection, otherwise 1 mark for triangle reflected but in wrong position]

4   a) Scale factor from *EFGH* to *ABCD* = $9 \div 6 = 1.5$ *[1 mark]*
        EF = $6 \div 1.5 = 4$ cm *[1 mark]*
        [2 marks available in total — as above]
    b) BC = $4 \times 1.5 = 6$ cm *[1 mark]*

5
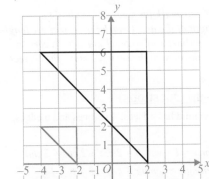

    [3 marks available — 3 marks for correct enlargement, otherwise 2 marks for a correct triangle but in the wrong position or for an enlargement from the correct centre but of the wrong scale factor, or 1 mark for 2 lines enlarged by the correct scale factor anywhere on the grid]

Page 114 (Warm-up Questions)

1   42 cm

2   a) area = length $\times$ width, A = l $\times$ w
    b) circumference = $\pi \times$ diameter, C = $\pi \times$ D (or C = 2πr)
    c) area = base $\times$ vertical height, A = b $\times$ h

3   10.5 m² *(Area = ½ $\times$ base $\times$ vertical height = 0.5 $\times$ 3 $\times$ 7)*

4   201.06 cm² to 2 d.p. (or 201.09 cm² to 2 d.p. using π = 3.142)
    (Area = $\pi r^2 = \pi \times 8^2$)

5 a) A straight line that just touches the outside of a circle.
   b)

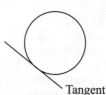

Tangent

Page 115 (Exam Questions)

3 a) Area of trapezium = ½(8 + 11) × 6
              = ½ × 19 × 6 = 57 cm²
   Area of triangle = area of trapezium ÷ 3 = 57 ÷ 3 = 19 cm²
   Total area of the shape = area of trapezium + area of triangle
              = 57 + 19 = 76 cm²
   [3 marks available — 1 mark for the area of the trapezium, 1 mark for the area of the triangle, 1 mark for correct final answer]
   b) Area of triangle = ½ × base × height
              19 = ½ × 8 × height *[1 mark]*
   height = 19 ÷ 4 = 4.75 cm *[1 mark]*
   [2 marks available in total — as above]

4 Area of rectangle = 6 × 8 = 48 cm² *[1 mark]*
   Base of triangle = 8 cm – 5 cm = 3 cm
   Height of triangle = 6 cm – 2 cm = 4 cm
              [1 mark for base and height]
   Area of triangle = $\frac{1}{2}$ × 3 × 4 = 6 cm² *[1 mark]*
   Area of shaded area = 48 – 6 = 42 cm² *[1 mark]*
   [4 marks available in total — as above]

5 Circumference of full circle = 2 × π × 6 = 12π cm
   Length of arc = $\frac{30}{360}$ × circumference of circle
   = $\frac{30}{360}$ × 12π = π cm
   Perimeter of sector = π + 6 + 6 = 15.1415... = 15.1 cm (3 s.f.)
   Area of full circle = π × 6² = 36π cm²
   Area of sector = $\frac{30}{360}$ × area of circle
   = $\frac{30}{360}$ × 36π = 3π cm² = 9.4247... = 9.42 cm² (3 s.f.)
   [5 marks available — 1 mark for a correct method for calculating the length of the arc, 1 mark for correct arc length, 1 mark for correct perimeter of sector, 1 mark for a correct method for finding the area of the sector, 1 mark for correct area of sector]

Page 122 (Warm-up Questions)

1 A triangle-based pyramid has 4 faces, 4 vertices and 6 edges.
2 4 cm
3 8 cm³ *(8 × 1 cm³, or v = l × w × h = 2 cm × 2 cm × 2 cm)*
4 672 cm³ *(area of triangle × length = ½ × 12 × 8 × 14)*
5 624π cm³
6 6 cm
7 a)   b)

   c)

Pages 123-124 (Exam Questions)

2 Split the shape into two cuboids, looking at the front elevation. The bottom cuboid has 4 × 2 × 4 = 32 cubes in it. The cuboid at the top has 2 × 2 × 4 = 16 cubes in it. So there are 32 + 16 = 48 cubes in the shape.
   [2 marks available — 1 mark for a correct calculation, 1 mark for the correct answer]
   You might have split your shape up differently — as long as your working is correct and you get the correct answer, you'll get all the marks.

3

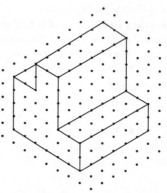

   [2 marks available — 2 marks for a correct diagram, otherwise 1 mark for the correct cross-section but wrong length]

4 Volume of sphere = $\frac{4}{3}\pi r^3$ = $\frac{4}{3}$ × π × 15³ *[1 mark]* = 4500π cm³
   = 14 137.166... = 14 100 cm³ (3 s.f.) *[1 mark]*
   [2 marks available in total — as above]

5 a) Volume = 90 × 40 × 30 *[1 mark]* = 108 000 cm³ *[1 mark]*
   [2 marks available in total — as above]
   b) Volume of cuboid = length × width × height
              108 000 = 120 × width × 18
              108 000 = 2160 × width
              width = 108 000 ÷ 2160 = 50 cm
   [2 marks available — 1 mark for correctly rearranging the formula to find the width, 1 mark for the correct answer]

6 a) Volume of water in paddling pool = π × r² × h
              = π × 100² × 40 *[1 mark]* = 400 000π cm³ *[1 mark]*
   [2 marks available in total — as above]
   b) Time it will take to fill to 40 cm = 400 000π ÷ 300 *[1 mark]*
              = 4188.790... seconds
   Convert to minutes = 4188.790... ÷ 60
              = 69.813... = 70 minutes (to the nearest minute) *[1 mark]*
   [2 marks available in total — as above]

Page 125 (Revision Questions)

1 H: 2 lines of symmetry, rotational symmetry order 2
   Z: 0 lines of symmetry, rotational symmetry order 2
   T: 1 line of symmetry, rotational symmetry order 1
   N: 0 lines of symmetry, rotational symmetry order 2
   E: 1 line of symmetry, rotational symmetry order 1
   ✕: 4 lines of symmetry, rotational symmetry order 4
   S: 0 lines of symmetry, rotational symmetry order 2

2 2 angles the same, 2 sides the same, 1 line of symmetry, no rotational symmetry.

3 2 lines of symmetry, rotational symmetry order 2

4 Congruent shapes are exactly the same size and same shape. Similar shapes are the same shape but different sizes.

5 a) D and G   b) C and F

6 $b = 89$, $y = 5$

7 a) Translation of $\binom{-2}{-4}$.

   b) Reflection in $x = 0$ (the y-axis).

8

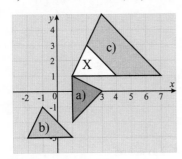

9 Area $= \frac{1}{2}(a + b) \times h$

10 36 cm²

11 52 cm²

12 Area = 153.94 cm² (2 d.p.)

   Circumference = 43.98 cm (2 d.p.)

13 E.g.

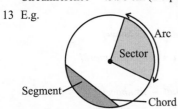

14 Area = 7.07 cm² (2 d.p.)

   Perimeter = 10.71 cm (2 d.p.)

15 a) faces = 5, edges = 8, vertices = 5

   b) faces = 2, edges = 1, vertices = 1

   c) faces = 5, edges = 9, vertices = 6

16 150 cm²

17 125.7 cm² (1 d.p.)

18 Volume $= \pi r^2 h$

19 360 cm³

20 a) 113.10 cm³ (2 d.p.)

   b) 4.5 s (1 d.p.)

21 Front: Side:

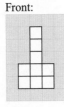

Plan:

Section Six — Angles and Geometry

Page 131 (Warm-up Questions)

1 a) Any angle less than 90°

   b) Any angle greater than 90° and less than 180°

   c) Any angle greater than 180° and less than 360°

   d) 90°

2 a) 37°   b) 287°

   Accept answers within 2°

3 a)   b)   c)

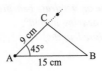

4 $x = 110°$, $y = 40°$

   Using angles in an isosceles triangle and angles on a straight line.

5 $a = 120°$, $b = 60°$

   Using the rule for allied angles (60° + a = 180°) and using the rule for corresponding angles (b = 60°).

6 540°

Page 132 (Exam Questions)

3 $a = 75°$ *[1 mark]*

   because vertically opposite angles are equal. *[1 mark]*

   [2 marks available in total — as above]

4 $70° + 90° + 97° = 257°$

   Angle $ADC = 360° - 257° = 103°$

   (angles in a quadrilateral add up to 360°)

   [2 marks available — 1 mark for a correct method, 1 mark for the correct answer]

5 Exterior angle $= 180° - 150° = 30°$ *[1 mark]*

   Number of sides $= 360° \div 30°$ *[1 mark]*

                 $= 12$ *[1 mark]*

   [3 marks available in total — as above]

6 Angle $BCG = $ Angle $CGE = x$ *(alternate angles)*

   So $78° + x = 180°$ *[1 mark]*

   $x = 180° - 78°$

   $x = 102°$ *[1 mark]*

   [2 marks available in total — as above]

   There are other ways to find x. For instance angles ACB and CGF are corresponding angles. You can then use angles on a straight line to find x.

7 The polygon is split into 5 triangles.

   Angles in a triangle add up to 180°

   Angles in polygon $= 5 \times 180°$

                        $= 900°$

   [3 marks available — 3 marks for correct explanation, otherwise 1 mark for stating angles in triangle add up to 180° and 1 mark for an attempt at adding to find angles in polygon]

Page 140 (Warm-up Questions)

1 a)

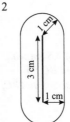

   b) 11 cm

2

3

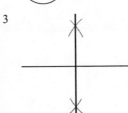

4 a) 1100 m (5.5 × 200)   b) 2.5 cm (500 ÷ 200)

5

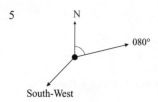

South-West

Pages 141-142 (Exam Questions)

3 a) Drawing of dining table is 4 cm long.
So 4 cm is equivalent to 2 m.
$2 \div 4 = 0.5$
Therefore scale is 1 cm to 0.5 m *[1 mark]*

   b) On drawing, dining table is 3 cm from shelves.
So real distance = $3 \times 0.5 = 1.5$ m *[1 mark]*

   c) The chair and the space around it would measure 4 cm × 5 cm
on the diagram and there are no spaces that big, so no, it
would not be possible.
*[2 marks available — 1 mark for correct answer, 1 mark for
reasoning referencing diagram or size of gaps available]*

4 260° (allow 258°-262°) *[1 mark]*
It's easier to measure the 100° angle and subtract it from 360°.

5 $180° - 79° = 101°$ *(allied angles)* *[1 mark]*
$360° - 101° = 259°$
Ishita travels on a bearing of 259°. *[1 mark]*
[2 marks available in total — as above]

6 Using the scale 1 cm = 100 m:
400 m = 4 cm and 500 m = 5 cm

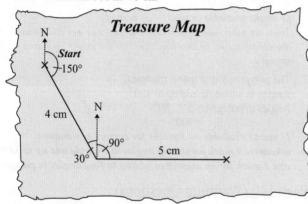

Treasure Map

*[3 marks available — 1 mark for line on accurate bearing
of 150°, 1 mark for line on accurate bearing of 090°,
1 mark for accurate 4 cm and 5 cm line lengths]*

7

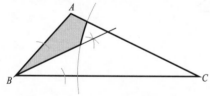

*[4 marks available — 1 mark for arc with radius of 6.5 cm with
centre at C, 1 mark for construction arcs on AB and BC for
angle bisector at ABC, 1 mark for correct angle bisector at ABC,
and 1 mark for the correct shading]*
Remember to leave in your construction lines.

Page 149 (Warm-up Questions)

1 7.6 m

2 1.94 cm

3 $\dfrac{12}{\sqrt{3}}$

4 $\dfrac{1}{2}$

5 a) $\begin{pmatrix} -15 \\ 9 \end{pmatrix}$ b) $\begin{pmatrix} -2 \\ -12 \end{pmatrix}$ c) $\begin{pmatrix} 8 \\ 15 \end{pmatrix}$

6 $\mathbf{b} - \mathbf{a}$ (or $-\mathbf{a} + \mathbf{b}$)

Page 150 (Exam Questions)

3 The triangle can be split into two right-angled triangles.

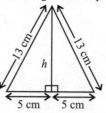

Let h be the height of the triangle:
$13^2 = 5^2 + h^2$ *[1 mark]*
$h^2 = 169 - 25 = 144$
$h = \sqrt{144}$ *[1 mark]*
$h = 12$ cm *[1 mark]*
[3 marks available in total — as above]

4 $\sin x = \dfrac{14}{18}$ *[1 mark]*

$x = \sin^{-1}\left(\dfrac{14}{18}\right)$

$x = 51.0575... = 51.1°$ (1 d.p) *[1 mark]*
[2 marks available in total — as above]

5 a) $\overrightarrow{AC} = \overrightarrow{AB} + \overrightarrow{BC}$ *[1 mark]*
$= 2\mathbf{c} + 2\mathbf{d}$ *[1 mark]*
[2 marks available in total — as above]

   b) $\overrightarrow{AL} = \dfrac{1}{2} \times \overrightarrow{AC} = \dfrac{1}{2} \times 2\mathbf{c} + \dfrac{1}{2} \times 2\mathbf{d}$ *[1 mark]*
$= \mathbf{c} + \mathbf{d}$ *[1 mark]*
[2 marks available in total — as above]

   c) $\overrightarrow{BL} = \overrightarrow{BA} + \overrightarrow{AL}$ *[1 mark]*
$= -2\mathbf{c} + \mathbf{c} + \mathbf{d}$
$= -\mathbf{c} + \mathbf{d}$ *[1 mark]*
[2 marks available in total — as above]

Pages 151-152 (Revision Questions)

1 An obtuse angle

2 360°

3 a) 154°   b) 112°   c) 58°

4 140°

5 60°

6 1080°

7 162°

8 (Not full size)

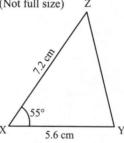

9 A circle

10

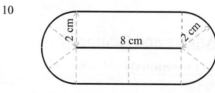

(Not full size)

11

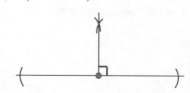

12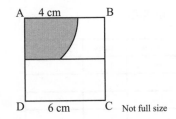

A   4 cm   B

D   6 cm   C   Not full size

13   295°

14   280°

15   26 km

16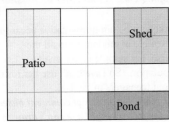

Shed

Patio

Pond

17

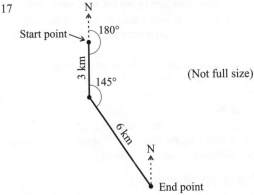

N

Start point   180°

3 km

145°

(Not full size)

6 km

N

End point

18   14.5 cm

19

$\dfrac{O}{S \times H}$ $\dfrac{A}{C \times H}$ $\dfrac{O}{T \times A}$

20   $x = 33.4°$ (1 d.p.)

21   $x = 51.8°$ (1 d.p.)

22   $12\sqrt{3}$ cm

23   $\tan 45° + \sin 60° = 1 + \dfrac{\sqrt{3}}{2}$

$= \dfrac{2}{2} + \dfrac{\sqrt{3}}{2}$

$= \dfrac{2 + \sqrt{3}}{2}$

24   a) $\begin{pmatrix} -3 \\ -8 \end{pmatrix}$ b) $\begin{pmatrix} 20 \\ -10 \end{pmatrix}$

25   $2\mathbf{a} + 3\mathbf{b}$

26   $5\mathbf{a} - 7\mathbf{b}$

27   $-\mathbf{i} + \mathbf{j}$

Section Seven — Probability and Statistics

Page 157 (Warm-up Questions)

1   $\dfrac{1}{3}$

2   $\dfrac{5}{6}$

3   a)   HH, HT, TH, TT

   b) $\dfrac{1}{2}$ or 0.5

4

Score	Relative frequency
1	0.14
2	0.137
3	0.138
4	0.259
5	0.161
6	0.165

5   $\dfrac{30}{120} = 0.25$

6   250

Pages 158-159 (Exam Questions)

3   $10 - 4 = 6$ red counters.

P(red) $= \dfrac{6}{10} = 0.6$

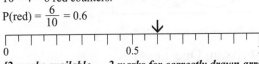

0   0.5   1

[2 marks available — 2 marks for correctly drawn arrow, otherwise 1 mark for finding the correct probability of picking a red counter]

You could also work out the probability of a blue counter (0.4) and subtract it from 1 to get the probability of a red counter.

4   a)   (Hockey, Netball), (Hockey, Choir), (Hockey, Orienteering), (Orchestra, Netball), (Orchestra, Choir), (Orchestra, Orienteering), (Drama, Netball), (Drama, Choir), (Drama, Orienteering).

   [2 marks available — 2 marks for listing all 9 correct combinations, otherwise 1 mark if at least 5 combinations are correct]

   b)   There are 9 combinations and 1 of them is hockey and netball, so P(hockey and netball) $= \dfrac{1}{9}$ *[1 mark]*

   c)   There are 9 combinations and 3 of them involve drama on Monday, so P(drama on Monday) $= \dfrac{3}{9} = \dfrac{1}{3}$ *[1 mark]*

   You could also count the choices for Monday — there are 3, and 1 of them is drama.

5   a)   EHM, EMH, HME, HEM, MEH, MHE

   [2 marks available — 2 marks for listing all 6 correct combinations, otherwise 1 mark if at least 3 combinations are correct]

   b)   There are 6 possible combinations and in 3 of them he does Maths before English (HME, MEH, MHE).

   So P(Maths before English) $= \dfrac{3}{6} = \dfrac{1}{2}$ *[1 mark]*

6   a)

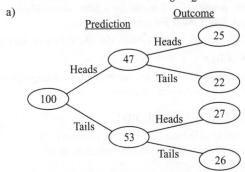

Prediction   Outcome

Heads   25

47

Heads   Tails   22

100

Heads   27

Tails   53

Tails   26

   [2 marks available — 1 mark for the correct numbers for the predictions, 1 mark for the correct numbers for the outcomes]

   b)   She predicted the flip correctly $25 + 26 = 51$ times out of 100 *[1 mark]*, so relative frequency $= \dfrac{51}{100} = 0.51$ *[1 mark]*

   [2 marks available in total — as above]

7   a)   Relative frequency of hitting the target with a left-handed throw $= \dfrac{12}{20} = \dfrac{3}{5}$ or 0.6.

   [2 marks available — 1 mark for a correct method, 1 mark for the correct answer]

   b)   E.g. The estimated probability is more reliable for her right hand because she threw the ball more times with that hand. *[1 mark]*

Page 163 (Warm-up Questions)

1   $\dfrac{1}{6} \times \dfrac{1}{6} \times \dfrac{1}{6} = \dfrac{1}{216}$

2   $0.5 \times 0.7 = 0.35$

3   a) $0.2 \times 0.25 = 0.05$ b) $0.2 + 0.25 = 0.45$

4   $\dfrac{6}{10} \times \dfrac{6}{10} = \dfrac{36}{100} = \dfrac{9}{25}$

5 a)

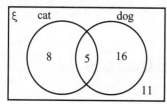

b) P(owns a cat but not a dog) = $\frac{8}{40} = \frac{1}{5}$

Page 164 (Exam Questions)

2 a) P(4 or 5) = P(4) + P(5)
= 0.25 + 0.1 *[1 mark]*
= 0.35 *[1 mark]*
[2 marks available in total — as above]

b) P(1 and 3) = P(1) × P(3)
= 0.3 × 0.2 *[1 mark]*
= 0.06 *[1 mark]*
[2 marks available in total — as above]

3 a) P(no prize) = 1 − 0.3 = 0.7 *[1 mark]*

b) P(no prize on either game) = P(no prize) × P(no prize)
= 0.7 × 0.7 *[1 mark]*
= 0.49 *[1 mark]*
[2 marks available in total — as above]

4 a) Number of students who only like apples = 70 − 20 = 50
Number of students who only like bananas = 40 − 20 = 20
Number of students who don't like either
= 100 − 50 − 20 − 20 = 10

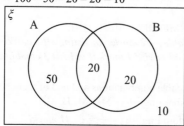

[3 marks available — 3 marks for a fully correct Venn diagram, lose 1 mark for each incorrect value]

b) 50 + 20 + 20 = 90 students out of 100 like apples or bananas,
so P(A ∪ B) = $\frac{90}{100} = \frac{9}{10}$ or 0.9 .
[2 marks available — 1 mark for 90 students liking either fruit, 1 mark for the correct answer.]

Page 168 (Warm-up Questions)

1 Two from, e.g: sample too small, one city centre not representative of the whole of Britain, only done in one particular place.

2 Continuous data

E.g.

Weight (w kg)	Tally	Frequency
$0 \le w < 10$		
$10 \le w < 20$		
$20 \le w < 30$		
$30 \le w < 40$		
$w \ge 40$		

3 a) This question is ambiguous. "A lot of television" can mean different things to different people.

b) This is a leading question, inviting the person to agree.

c) The answers to this question do not cover all possible options.

4 First, order the numbers:
−14, −12, −5, −5, 0, 1, 3, 6, 7, 8, 10, 14, 18, 23, 25
Mean = 5.27 (2 d.p.), Median = 6, Mode = −5, Range = 39

5 2, 4 and 6

Page 169 (Exam Questions)

2 a) E.g.

Number of chocolate bars	Tally	Frequency
0-2		
3-5		
6-8		
9-11		
12 or more		

[2 marks available — 1 mark for a suitable tally table, 1 mark for non-overlapping classes that cover all possible values]

b) E.g. Faye's results are likely to be unrepresentative because she hasn't selected her sample at random from all the teenagers in the UK. Also, her sample is too small to represent the whole population. So Faye can't use her results to draw conclusions about teenagers in the UK.
[2 marks available — 1 mark for a correct comment based on sample size, 1 mark for stating that Faye can't draw conclusions about teenagers in the UK with reasoning]

3 a) Yes, the mean number is higher than 17 because the 11th data value is higher than the mean of the original 10 values.
[1 mark]

b) You can't tell if the median number is higher than 15, because you don't know the other data values.
[1 mark]

4 a) Mode = 1 *[1 mark]*

b) Median = (25 + 1) ÷ 2 = 13th value *[1 mark]*.
13th value is shown by the 2nd bar, so median = 1 *[1 mark]*.
[2 marks available in total — as above]

Page 174 (Warm-up Questions)

1 a)

Red	⬤⬤⬤⬤⬤
Purple	⬤⬤
Blue	⬤⬤⬤⬤◗
Orange	⬤⬤⬤◗

b) No. According to the pictogram, there are only (4 × 8) + (½ × 8) = 32 + 4 = 36 blue balloons at the shop, and Ellen needs 37.

2

	Walk	Car	Bus	Total
Male	15	21	13	49
Female	18	11	22	51
Total	33	32	35	100

3 Giraffe = 31 × 4° = 124°
Penguin = 25 × 4° = 100°
Lemur = 34 × 4° = 136°

4 a) There is a strong positive correlation. The taller the student, the bigger their shoe size.

b) Approximately 7.5

c) The estimate should be reliable because **[either]** 1.75 m is within the range of the known data **[or]** the graph shows strong correlation.

d) Yes. If you plotted Amal's height and shoe size, it would be close to the other points on the graph, so it follows the trend.

Page 175 (Exam Questions)

2 E.g.

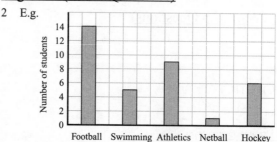

[4 marks available — 1 mark for a suitable scale starting from zero on the vertical axis, 1 mark for correctly labelling the vertical axis, 1 mark for bars of equal width and all bar heights correct, 1 mark for all bars correctly labelled]

3 a) $15:20 = 3:4$ *[1 mark]*

b) $\frac{20}{50} \times 100$ *[1 mark]* $= 40\%$ *[1 mark]*
    [2 marks available in total — as above]

4 a)

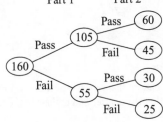

[1 mark for both points plotted correctly]

b) $\frac{5}{14}$ *[1 mark]*

c) E.g. In general, as the height increases, the weight also increases — there is a positive correlation.
    [1 mark for any answer indicating a positive correlation]

Page 180 (Warm-up Questions)

1 a) Mode = 3   b) Median = 3

2 E.g.

Height (h cm)	Tally	Frequency
$150 < h \leq 160$	||	2
$160 < h \leq 170$	||||	5
$170 < h \leq 180$	|||	3
$180 < h \leq 190$	||	2

3 a) Estimated range = $185 - 145 = 40$

b) Median is in the group containing the 40th value, so the median group is $155 \leq x < 165$.

c) Modal Group = $165 \leq x < 175$

4 The mode is 30°C but most of the temperatures are less than 20°C, so it's not representative of all the data.

Page 181 (Exam Questions)

2 a) Total number of people = $12 + 18 + 9 + 21 = 60$
       Multiplier = $360 \div 60 = 6$
       Plain: $12 \times 6 = 72°$
       Salted: $18 \times 6 = 108°$
       Sugared: $9 \times 6 = 54°$
       Toffee: $21 \times 6 = 126°$

       Plain (72°)
       Toffee (126°)
       Salted (108°)
       Sugared (54°)

    [4 marks available — 1 mark for one sector correctly drawn, 1 mark for a second sector correctly drawn, 1 mark for a complete pie chart with all angles correct, 1 mark for correct labels]

b) E.g. Kim is not right because there is no information about the number of people in the ice-cream survey. *[1 mark]*

3 a) Max value = 63 mm, min value = 8 mm *[1 mark for both]*, so range = $63 - 8 = 55$ mm *[1 mark]*
       E.g. a range of 55 mm isn't a good reflection of the spread of the data because most of the data is much closer together.
       [1 mark for a correct comment]
       [3 marks available in total — as above]
       Or you could say that the single value of 63 mm has a big effect on increasing the value of the range so that it doesn't represent the spread of the rest of the data.

b) Median rainfall in June = $(12 + 1) \div 2 = 6.5$th value
    $= (29 + 30) \div 2 = 29.5$ mm
    E.g. The rainfall was generally higher in June, as the median was higher. The rainfall in June was much more varied than in November as the range was much bigger.
    [3 marks available — 1 mark for calculating the median rainfall in June, 1 mark for a correct statement comparing the medians and 1 mark for a correct statement comparing the ranges]

Page 182 (Revision Questions)

1 $\frac{8}{50} = \frac{4}{25}$

2 0.7

3 a) HHH, HHT, HTH, THH, TTH, HTT, THT, TTT   b) $\frac{3}{8}$

4 a) See page 155   b) See page 156

5 a)

       Part 1   Part 2
                       Pass ── 60
            Pass ── 105
                       Fail ── 45
       160
                       Pass ── 30
            Fail ── 55
                       Fail ── 25

b) Relative frequency of:
    pass, pass $= \frac{60}{160} = \frac{3}{8}$ or 0.375
    pass, fail $= \frac{45}{160} = \frac{9}{32}$ or 0.28125
    fail, pass $= \frac{30}{160} = \frac{3}{16}$ or 0.1875
    fail, fail $= \frac{25}{160} = \frac{5}{32}$ or 0.15625

c) $300 \times 0.375 = 113$ people (to the nearest whole number)

6 0.5

7

                                2nd card
                        $\frac{1}{4}$ ── Heart
         1st card
                 Heart
        $\frac{1}{4}$ $\frac{3}{4}$ ── No heart
                        $\frac{1}{4}$ ── Heart
        $\frac{3}{4}$
                 No heart
                        $\frac{3}{4}$ ── No heart

    P(no hearts) $= \frac{9}{16}$

8 a)

       ξ Tea   Coffee
            14   20   30
       36

b) $\frac{64}{100} = \frac{16}{25}$

9 A sample is part of a population. Samples need to be representative so that conclusions drawn from sample data can be applied to the whole population.

10 Qualitative data

11

Pet	Tally	Frequency
Cat	|||| |||	8
Dog	|||| |	6
Rabbit	||||	4
Fish	||	2

12 Mode = 31, Median = 24, Mean = 22, Range = 39

13 Count the number of symbols, then use the key to work out what frequency they represent.

14 E.g.

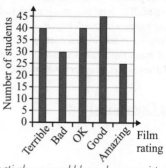

Alternatively, you could have drawn a pictogram.

15 a)

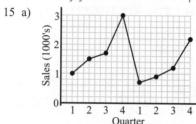

b) There is a seasonal pattern that repeats itself every 4 points. Sales are lowest in the first quarter and highest in the fourth quarter.

16 Draw a pie chart to show the proportions.

17 a) 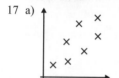 b) c)

18 a) Modal class is: $1.5 \leq y < 1.6$.

b) Class containing median is: $1.5 \leq y < 1.6$

c) Estimated mean = 1.58 m (to 2 d.p.)

19 Outliers can have a big effect on increasing or decreasing the value of the mean or range, so that it doesn't represent the rest of the data set very well.

20 Black cars were only owned by men and silver cars were only owned by women. So black cars were more popular amongst men and silver cars were more popular amongst women.
There are similar proportions of men and women owning blue and green cars. So blue and green cars are equally popular amongst men and women.
The proportion of men owning red cars was nearly double the proportion of women owning red cards. So red cars were almost twice as popular amongst men as women.

Practice Paper 1

1 $\frac{113}{1000}$ *[1 mark]*

2 $40 : 25 = 8 : 5$ *[1 mark]*

3 a) F *[1 mark]*

b) $x = 3$ *[1 mark]*

4 $3.97 \times 1000 = 3970$ m *[1 mark]*

5 a) $-12, -8, -6, 2, 6$ *[1 mark]*

b) $2 - -8 = 10$ *[1 mark]*

6 $5p \times 28 = 140p$
$10p \times 41 = 410p$
$140p + 410p = 550p$
$550p \div 50p = 11$, so she gets 11 50p coins.
[3 marks available — 1 mark for the total of the 5p or 10p coins, 1 mark for the overall total, 1 mark for the final answer]

7 $1.2 - 0.2 \times 4 = 1.2 - 0.8 = 0.4$ *[1 mark]*
So $\frac{1.2 - 0.2 \times 4}{0.05} = \frac{0.4}{0.05} = \frac{40}{5} = 8$ *[1 mark]*
[2 marks available in total — as above]

8 E.g. 6 (multiple of 3) + 6 (multiple of 6)
= 12 (not a multiple of 9) *[1 mark]*

9 a) $9a + 7b$ *[1 mark]*

b) $6a^2$ *[1 mark]*

10 1% of £300 = £3, so 2% = £3 × 2 = £6 *[1 mark]*
Interest for 4 years = £6 × 4 = £24 *[1 mark]*
£300 + £24 = £324 *[1 mark]*
[3 marks available in total — as above]

11 a) $3 + 6 + 5 + 2 + 6 + 1 + 3 = 26$ pupils
The frequency of the final bar should be $30 - 26 = 4$

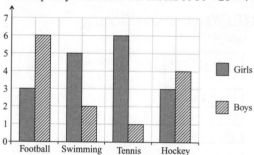

[2 marks available — 2 marks for correctly drawing missing bar, otherwise 1 mark for attempting to add the seven frequencies shown on the chart]

b) Number of children whose favourite sport is swimming is $5 + 2 = 7$. The probability is $\frac{7}{30}$
[2 marks available — 1 mark for the numerator (7) and 1 mark for the denominator (30)]

c) 2 boys chose swimming and 6 girls chose tennis, so the ratio is $2 : 6 = 1 : 3$ in its simplest form.
[2 marks available — 2 marks for the correct answer, otherwise 1 mark for 2 : 6]

12 Geometric *[1 mark]*
Rule: multiply the previous term by 4 *[1 mark]*

13 a)

Pattern	Number of triangles	Number of dots	Number of lines
Pattern 1	1	3	3
Pattern 2	2	4	5
Pattern 3	3	5	7
Pattern 4	4	6	9

[1 mark]

b) The number of lines column increases by 2 each time. The column would continue 11, 13, 15, 17, 19, 21. Pattern 10 has 21 lines.
[2 marks available — 1 mark for finding the sequence, 1 mark for the correct number of lines in Pattern 10]
You could also find an expression for the nth term (2n + 1) and work out the value when put n = 10, (2 × 10) + 1 = 21.

c) (i) The number of dots is two more than the pattern number, so $D = n + 2$
[2 marks available — 2 marks for D = n + 2, otherwise 1 mark for n + 2, D − n = 2, or '2 more than the pattern number']

(ii) Number of dots in pattern 200 = 200 + 2 = 202 *[1 mark]*

14 The number of tickets sold was:
Adults: $(8 \times 3) + 6 = 30$
Child: $(8 \times 5) + 2 = 42$
Senior: $(8 \times 2) + 4 = 20$
Adult tickets: $30 \times £9 = £270$
Child tickets: $42 \times £5 = £210$
Senior tickets: $20 \times £6.50 = £130$
Total sales = $270 + 210 + 130 = £610$
[6 marks available — 1 mark for each correct number of tickets sold (adult, child, senior), 1 mark for multiplying number of tickets sold by cost, 1 mark for attempting to add up the total sales, 1 mark for correct answer]

15 a) 067° (Accept answer between 065° and 069°) *[1 mark]*

b) Distance on map = 4.3 cm
Actual distance = $4.3 \times 100 = 430$ metres
[2 marks available — 2 marks for an answer in the range 420 m-440 m, otherwise 1 mark for a measurement in the range 4.2 cm-4.4 cm]

c) Draw an arc of a circle with radius 4.5 cm from the house and 7 cm from the greenhouse.

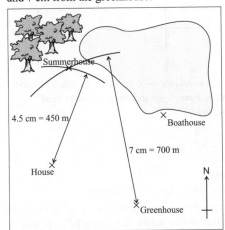

[2 marks available — 2 marks for summerhouse plotted in the correct position with construction arcs shown, otherwise 1 mark for one correct arc or for the summerhouse in the correct position but construction arcs not shown]

16 The top tier needs 50% of 800 g = 400 g of sultanas *[1 mark]*
The middle tier needs 75% of 800 g = 600 g of sultanas *[1 mark]*
One wedding cake needs $800 + 400 + 600 = 1800$ g of sultanas *[1 mark]*
Five wedding cakes need 1800 g $\times 5 = 9000$ g $= 9$ kg of sultanas *[1 mark]*, so Angie does not have enough sultanas. *[1 mark]*
[5 marks available in total — as above]

17 a) 69p rounds to 70p and 2.785 kg rounds to 3 kg to 1 s.f.
So estimate is $70p \times 3$ *[1 mark]*
$= 210p$ *[1 mark]*
[2 marks available in total — as above]
If you've done 70p × 2.8 = 196p you'll still get the marks.

b) E.g. The estimate is bigger than the actual cost, as both numbers were rounded up. *[1 mark]*

18 a) Read off the graph: £300 = $540 *[1 mark]*

b) After spending $390, he is left with $150 *[1 mark]*
Convert to yuan: $150 \times 6 = 900$ yuan *[1 mark]*
[2 marks available in total — as above]

19 $3^{-2} = \dfrac{1}{9}$ *[1 mark]*
$k \times \dfrac{1}{9} = 4$, so $k = 4 \times 9 = 36$ *[1 mark]*
[2 marks available in total — as above]

20 Angles in a square are 90°.
There are three squares around O, so $90° \times 3 = 270°$. *[1 mark]*
Angles round a point add to 360°,
so angle $AOB = (360 - 270) \div 3 = 30°$ *[1 mark]*
The two base angles in an isosceles triangle are equal,
so angle $OAB = (180 - 30) \div 2$ *[1 mark]*
$= 75°$ *[1 mark]*
[4 marks available in total — as above]

21 $1\dfrac{2}{3} \times 1\dfrac{5}{8} = \dfrac{5}{3} \times \dfrac{13}{8} = \dfrac{65}{24} = 2\dfrac{17}{24}$
[3 marks available – 1 mark for converting both numbers to improper fractions, 1 mark for multiplying, 1 mark for the correct answer]

22 $594\,000\,000\,000 = 5.94 \times 10^{11}$ *[1 mark]*

23 Year 9 — $\dfrac{9}{20} = \dfrac{45}{100} = 45\%$
Year 10 — 49%
Year 11 — $\dfrac{12}{12 + 13} = \dfrac{12}{25} = \dfrac{48}{100} = 48\%$
So Year 10 has the largest proportion of girls.
[3 marks available — 2 marks for converting two proportions to the same form as the third, and 1 mark for the correct answer.]
You could convert any two proportions to the form of the third, e.g. convert Year 10 and Year 11 into fractions.

24 Shape A: Area = $\pi \times 4^2 = 16\pi$ cm²
Shape B: Area = $\dfrac{80}{360} \times \pi \times 6^2 = 8\pi$ cm²
$16\pi = 2 \times 8\pi$ so the area of A is twice the area of B.
[4 marks available — 1 mark for the area of shape A, 1 mark for the correct method to find the area of shape B, 1 mark for the area of shape B, 1 mark for showing that the area of shape A is twice the area of shape B]

25 a) $AC^2 = AD^2 + DC^2 = 3^2 + 4^2 = 9 + 16 = 25$
So $AC = \sqrt{25} = 5$ cm
$x^2 = AC^2 + AB^2 = 5^2 + 12^2 = 25 + 144 = 169$
So $x = \sqrt{169} = 13$ cm
[4 marks available — 1 mark for using Pythagoras' theorem to find the length of AC, 1 mark for the correct length of AC, 1 mark for using Pythagoras' theorem to find the length of x, 1 mark for the correct length of x]

b) Area of triangle $ACD = 0.5 \times 4 \times 3 = 6$ cm²
Area of triangle $ABC = 0.5 \times 5 \times 12 = 30$ cm²
Area of quadrilateral $ABCD = 6 + 30 = 36$ cm²
[2 marks available — 1 mark for finding the areas of triangles ACD and ABC, 1 mark for adding to find the area of the quadrilateral]

26 $2x + y = 10 \xrightarrow{\times 2} 4x + 2y = 20$ *[1 mark]*

$\begin{aligned} 4x + 2y &= 20 \\ -\ 3x + 2y &= 17 \\ \hline x &= 3 \end{aligned}$ *[1 mark]*

$4(3) + 2y = 20$
$2y = 20 - 12$
$2y = 8$
$y = 4$ *[1 mark]*

[3 marks available in total — as above]
You could have found the value of y first then used it to find x.

Practice Paper 2

1 $\dfrac{3}{5} = 0.6 = 60\%$ *[1 mark]*

2 $20 + 4 = 24$
$24 \div 2 = 12$ *[1 mark]*

3
Menu Item	Number Ordered	Cost per Item	Total
Tea	2	£1.25	£2.50
Coffee	6 *[1 mark]*	£1.60	£9.60
Cake	4	£1.30 *[1 mark]*	£5.20
Tip			£2.50
		Total cost	£19.80 *[1 mark]*

[3 marks available in total — as above]

4 a)

cone — 4
sphere — 3
cylinder — 2
— 1

[2 marks available — 2 marks for all lines correct, otherwise 1 mark for 2 lines correct]

b) 6 *[1 mark]*

5 a) 107.3158498 *[1 mark]*
Your calculator may display more or fewer digits than this.

b) 107.32 *[1 mark]*
If you got a) wrong but rounded it correctly, you'll still get the mark for part b).

6 The possible numbers are:

    4356   4536
    5346   5364
    5436   5634
    6354   6534

[2 marks available — 2 marks if all 8 possible numbers are given with no errors or repetitions, otherwise 1 mark if at least 5 of the possible numbers are listed]

7 64 (4^3 and 8^2) *[1 mark]*

8 a) $4(a + 2) = 4a + 8$ *[1 mark]*
b) $y^2 + 5y = y(y + 5)$ *[1 mark]*

9 a) Mode is the most common value = 6 years *[1 mark]*
b) Rewrite data in order: 5, 6, 6, 7, 9, 11, 12
Median is the middle (4th) value = 7 years *[1 mark]*
c) Mean = $\dfrac{6+12+9+6+5+7+11}{7}$ *[1 mark]*
$= \dfrac{56}{7} = 8$ years *[1 mark]*
[2 marks available in total — as above]

10 a) Two equilateral triangles join together to form a rhombus. So 2 lines of symmetry. *[1 mark]*
b) E.g.

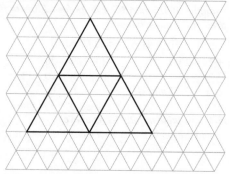

[1 mark]

11 185% of £3500 = 1.85 × 3500 *[1 mark]*
= £6475 *[1 mark]*
[2 marks available in total — as above]
You could also build up to 185%, by finding 50%, 5% etc. and adding these to the original value.

12 Method 1: 2 × 27 = 54 miles
           Expenses = 54 × 0.40 = £21.60 *[1 mark]*
Method 2: Expenses = (8 × 0.40) + 17.60 = £20.80 *[1 mark]*
She should travel by the cheaper method, which is Method 2 (by car and train). *[1 mark]*
[3 marks available in total — as above]

13 There are $\dfrac{3}{4}$ × 28 = 21 red cars and
$\dfrac{3}{8}$ × 16 = 6 red motorbikes *[1 mark for both]*.
So there are 21 + 6 = 27 red vehicles in total *[1 mark]*.
The probability of picking a car from the red vehicles
is $\dfrac{21}{27} = \dfrac{7}{9}$ *[1 mark]*.
[3 marks available in total — as above]

14 a)

x	−2	−1	0	1	2	3	4
y	11	9	7	5	3	1	−1

[2 marks available — 2 marks for all entries correct, otherwise 1 mark for at least 2 correct entries]

b)

[graph showing a straight line decreasing from (−2, 11) to (4, −1)]

[2 marks available — 1 mark for plotting at least 3 points correctly, 1 mark for a correct straight line]

c) −2 *[1 mark]*

15 $3(2x - 4) = 2x + 8$
$6x - 12 = 2x + 8$ *[1 mark]*
$4x = 20$ *[1 mark]*
$x = 5$ *[1 mark]*
[3 marks available in total — as above]

16 He has multiplied the denominator and numerator by 5, but he should have just multiplied the numerator by 5. *[1 mark]*

17 E.g. The vertical scale does not begin at 0 so the graph could be misinterpreted.
The horizontal scale does not increase by even amounts.
The graph is difficult to read due to the thickness of the line.
[3 marks available — 1 mark for each correct comment]

18 Angle $EBF = 90°$ *[1 mark]*
Angles on a straight line add up to 180°,
so angle $ABE = 180 - 90 - 39 = 51°$ *[1 mark]*
Angles ABE and DEH are corresponding angles,
so angle $DEH = 51°$ *[1 mark]*
[3 marks available in total – as above]

19 a) $L = 2(3x + 1) + 2(2x - 3) - 2$ *[1 mark]*
$= 6x + 2 + 4x - 6 - 2 = 10x - 6$ *[1 mark]*
[2 marks available in total — as above]
b) $L = 10x - 6 = 2(5x - 3)$, so it is always even as it can be written as 2 × a whole number, where the whole number is $(5x - 3)$
[2 marks available — 1 mark for writing the expression as $2(5x - 3)$ and 1 mark for explaining why this is always an even number.]

20

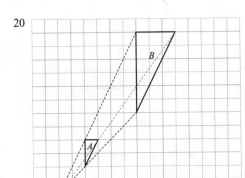

[2 marks available — 2 marks for image totally correct, otherwise 1 mark for 2 vertices in the correct position or for an image of the correct size but positioned incorrectly on the grid]

21 Elements of A are 3, 4, 5, 6
Elements of B are 1, 2, 3, 4, 6

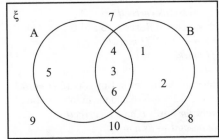

[3 marks available — 3 marks for a completely correct diagram, otherwise 1 mark for identifying the elements of sets A and B and 1 mark for correct elements in the intersection.]

22 8 litres of orangeade costs $(3 \times £1.60) + (5 \times £1.20) = £10.80$
So 1 litre of orangeade costs $£10.80 \div 8 = £1.35$
18 litres of orangeade cost $£1.35 \times 18 = £24.30$
[4 marks available — 1 mark for using the ratio to find the cost of 8 litres of orangeade, 1 mark for dividing by 8, 1 mark for multiplying by 18 and 1 mark for the correct final answer.]
There are several different methods you could use here. Any correct method with full working shown and a correct final answer would get 4 marks.

23 $y = \frac{x^2 - 2}{5}$
$5y = x^2 - 2$ *[1 mark]*
$5y + 2 = x^2$
$x = \pm\sqrt{5y + 2}$ *[1 mark]*
[2 marks available in total— as above]

24 a) 125 kW *[1 mark]*
   b) See diagram in part d).
      [1 mark for circling the point shown]
   c) Strong positive correlation *[1 mark]*
   d) Ignore the outlier when drawing a line of best fit.

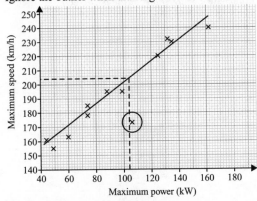

Maximum speed = 204 km/h (allow ± 2).
[2 marks available — 1 mark for drawing a line of best fit (ignoring the outlier), 1 mark for accurately reading from your graph the speed corresponding to a power of 104 kW]

e) 190 kW lies outside of the range of data plotted on the scatter graph. *[1 mark]*

25 Ollie's expression:
$(x + 4)^2 - 1 = (x + 4)(x + 4) - 1 = x^2 + 4x + 4x + 16 - 1$ *[1 mark]*
$= x^2 + 8x + 15$ *[1 mark]*

Amie's expression:
$(x + 3)(x + 5) = x^2 + 3x + 5x + 15 = x^2 + 8x + 15$, *[1 mark]*
so the two expressions are equivalent.
[3 marks available in total — as above]

26 a) Angle for car = $360 - 162 - 36 - 45 = 117°$ *[1 mark]*
No. of office assistants that travel by car is $= \frac{117}{360} \times 80$
$= 26$ *[1 mark]*
[2 marks available in total — as above]
   b) Total number of staff that travel by car
   $= 26 + 18 = 44$ *[1 mark]*
   44 represents 40% of the staff in the company, so the total number of staff (100%) is $44 \div 40 \times 100 = 110$ *[1 mark]*
   [2 marks available in total — as above]

27 a) Mass of cylinder = volume × density
   $= 1180 \times 2.7 = 3186$ g
   [2 marks available — 1 mark for using the density formula correctly and 1 mark for the correct final answer.]
   b) Mass of cube = mass of cylinder = 3186 g
   Volume of cube = mass ÷ density
   $= 3186 \div 10.5 = 303.428...$ cm³
   Side length $= \sqrt[3]{303.428...}$
   $= 6.719...$ cm = 6.7 cm (to 2 s.f.)
   [4 marks available — 1 mark for using the density formula correctly, 1 mark for finding the volume of the cube, 1 mark for attempting to find the cube root of the volume and 1 mark for the correct final answer.]

28 $\sin x = \frac{18}{24}$ *[1 mark]*
$x = \sin^{-1}\left(\frac{18}{24}\right) = 48.590... = 48.6$ (to 1 d.p.) *[1 mark]*
[2 marks available in total — as above]

Practice Paper 3

1 $0.4 > 0.34$
$\frac{3}{4} = 0.75$
$7\% < 0.7$
[2 marks available — 2 marks for all correct signs, otherwise 1 mark for 2 signs correct]

2 0.404 *[1 mark]*

3 20 800 *[1 mark]*

4 −16.2 *[1 mark]*

5 8 *[1 mark]*

6 a)

Result of exam	Tally	Frequency
Fail	IIII	4
Pass	HHH HHH I	11
Merit	HHH II	7
Distinction	II	2
	TOTAL:	24

[2 marks available — 1 mark for tally marks fully correct, 1 mark for correct frequencies]
   b) $\frac{4}{24} = \frac{1}{6}$
   [2 marks available — 2 marks for correct answer, otherwise 1 mark for $\frac{4}{24}$ or $\frac{2}{12}$]

240

c) E.g.

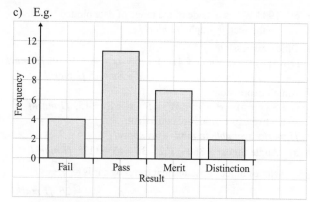

[3 marks available — 1 mark for correct vertical axis scale and labelling, 1 mark for 4 labelled bars of equal widths, 1 mark if the bars all have heights that agree with the frequencies found in (a)]

7 a) Impossible *[1 mark]*

b) There are 5 blue pencils, so evens. *[1 mark]*

8 a) Trapezium *[1 mark]*

b) Parallelogram *[1 mark]*

9 $\begin{pmatrix} -5 \\ 0 \end{pmatrix}$ *[1 mark]*

10 Square numbers: 1, 4, 9, 16, 25, 36...
Prime numbers: 2, 3, 5, 7, 11, 13, 17...
$2 \times 5 \times 9 = 90$, so Rick's numbers are 2, 5, and 9.
[3 marks available — 3 marks for correct answer, otherwise 1 mark for listing at least 2 square numbers or 2 prime numbers, 1 mark for 3 numbers which multiply to give 90, but which aren't 1 square number and 2 prime numbers.]
You could also answer this by using a factor tree to find that
$90 = 2 \times 3 \times 3 \times 5 = 2 \times 9 \times 5$

11 1 cupcake needs $132 \div 12 = 11$ g sugar.
So 30 cupcakes need $30 \times 11 = 330$ g sugar.
[2 marks available — 1 mark for a correct method, 1 mark for the correct answer]

12 Multiples of 12: 12, 24, 36, 48, 60 ...
Multiples of 10: 10, 20, 30, 40, 50, 60, ...
The smallest number they could each have bought is 60.
This is 6 packets of shortbread biscuits.
[3 marks available — 3 marks for the correct answer, otherwise 2 marks for finding that smallest number of biscuits they can buy is 60, or 1 mark for listing at least three multiples of 12 or 10]

13 Mary needs $2 \times 30 = 60$ scones, 30 tubs of clotted cream and 30 small pots of jam. *[1 mark]*
$60 \div 10 = 6$ packs of scones, costing $6 \times £1.35 = £8.10$ *[1 mark]*
$30 \div 6 = 5$ tubs of clotted cream,
costing $5 \times £2.95 = £14.75$ *[1 mark]*
30 pots of jam costing $30 \times £0.40 = £12$ *[1 mark]*
The total cost is $£8.10 + £14.75 + £12 = £34.85$,
so yes, she has enough money. *[1 mark]*
[5 marks available in total — as above]

14

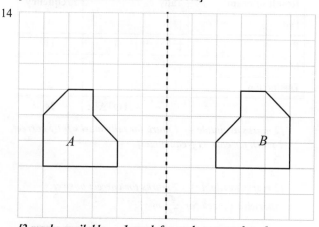

[2 marks available — 1 mark for each correct shape]

15

Plan view   Front elevation

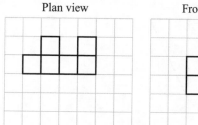

[2 marks available — 1 mark for each correct drawing]

16 Each trapezium has these dimensions:
Parallel sides: 12 cm and 7 cm
Height: 6 cm
So area $= \frac{1}{2}(12 + 7) \times 6$ *[1 mark]*
$= 9.5 \times 6 = 57$ cm² *[1 mark]*
[2 marks available in total — as above]

17 Box B contains $c + 4$ chocolates
Box C contains $2(c + 4) = 2c + 8$ chocolates
[1 mark for both expressions correct]
Total number of chocolates in Boxes A, B and C
$= c + c + 4 + 2c + 8 = 4c + 12$ *[1 mark]*
So $4c + 12 = 60$ *[1 mark]*
$4c = 48$
$c = 12$, so the number of chocolates in Box A is 12 *[1 mark]*,
the number in Box B is $12 + 4 = 16$ and the number in Box C is $2 \times 16 = 32$. *[1 mark for boxes B and C correct]*
[5 marks available in total — as above]

18 a) $y \times y \times y = y^3$ *[1 mark]*

b) Subtract the powers: $n^6 \div n^2 = n^4$ *[1 mark]*

c) Multiply the powers: $(a^4)^3 = a^{12}$ *[1 mark]*

19 Use Pythagoras' theorem on the triangle formed by one step and the ramp. Let x be the length of the ramp.
$20^2 + 55^2 = x^2$ *[1 mark]*
$x^2 = 3425$
So, $x = \sqrt{3425}$ cm *[1 mark]*
Total length of the ramp $= 2 \times \sqrt{3425}$ cm
$= 117.046...$ cm
$= 117$ cm (3 s.f.) *[1 mark]*
[3 marks available in total — as above]

20 a) Dave's speed is $180 \div 3 = 60$ km/h *[1 mark]*
Olivia's speed is $60 + 15 = 75$ km/h
So it takes her $180 \div 75 = 2.4$ hours *[1 mark]*
$= 2$ hours 24 minutes *[1 mark]*
[3 marks available in total — as above]

b) E.g. If Olivia drove a different distance at the same average speed, her journey time would be different *[1 mark]*.

21 a)

x	−3	−2	−1	0	1	2	3
y	4	0	−2	−2	0	4	10

[2 marks available — 2 marks for all three values correct, otherwise 1 mark if one value is correct]

Answers

b)

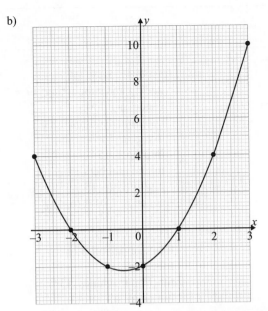

[2 marks available — 1 mark for plotting the points from your table and 1 mark for joining with a smooth curve to form a U-shaped curve]

22 2 + 3 + 5 = 10 'parts' in the ratio. Angles in a triangle add to 180°, so 10 parts = 180° and 1 part = 18°.
So, the three angles in the triangle are (2 × 18°) = 36°, (3 × 18°) = 54° and (5 × 18°) = 90°.
One angle is 90°, so the triangle is a right-angled triangle.
[3 marks available — 1 mark for finding the size of one part of the ratio, 1 mark for finding the size of at least one angle in the triangle and 1 mark for showing that one angle is 90°.]

23 a) 25% higher than £120 000 = 1.25 × £120 000
= £150 000, so House 4 *[1 mark]*

b) £161 280 – £144 000 = £17 280 *[1 mark]*
% change = $\frac{17\,280}{144\,000}$ × 100 *[1 mark]* = 12% *[1 mark]*

[3 marks available in total — as above]

24 Work back in Anna's sequence to find her 1st term:
17 – 3 – 3 – 3 = 8 *[1 mark]*
So the 1st term of Carl's sequence is 8 ÷ 2 = 4 *[1 mark]*
So the first five terms of Carl's sequence are 4, 10, 16, 22, 28.
His 5th term is 28. *[1 mark]*
[3 marks available in total — as above]

25 a) $x^2 + 7x – 18 = (x + 9)(x – 2)$
[2 marks available — 2 marks for the correct factorisation, otherwise 1 mark for an answer of the form (x ± a)(x ± b) where a and b are numbers that multiply to make 18]

b) $(x + 9)(x – 2) = 0$
$x = –9$ or $x = 2$
[1 mark for correct solutions using factorisation in (a)]

26 a)

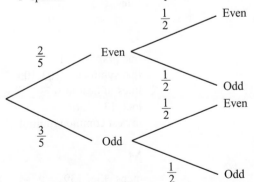

[2 marks available — 1 mark for the correct probabilities for the first spinner, 1 mark for the correct probabilities for the second spinner]

b) P(two odd numbers) = $\frac{3}{5} \times \frac{1}{2}$ *[1 mark]*
= $\frac{3}{10}$ *[1 mark]*
[2 marks available in total — as above]

27 Change in y = 8 – (–7) = 15
Change in x = 3 – (–2) = 5
So gradient = 15 ÷ 5 = 3
So y = mx + c becomes y = 3x + c
Put in x = 3 and y = 8 to find the value of c:
8 = 3(3) + c, which means c = 8 – 9 = –1
The equation of the line is y = 3x – 1.
[4 marks available — 1 mark for a correct method for finding the gradient, 1 mark for the correct gradient, 1 mark for putting one point into the equation, 1 mark for the correct answer]

28

Weight (w g)	Frequency	Mid-interval value	Frequency × mid-interval value
800 ≤ w < 1000	5	900	4500
1000 ≤ w < 1200	8	1100	8800
1200 ≤ w < 1400	9	1300	11 700
1400 ≤ w < 1600	3	1500	4500
Total	25	—	29 500

Estimate of mean 29 500 ÷ 25 = 1180 g.
[3 marks available — 1 mark for attempting to find the mid-interval and frequency × mid-interval values, 1 mark for dividing the sum of the frequency × mid-interval values by 25, 1 mark for the correct answer.]

Index

Index

Formulas in the Exams

GCSE Maths uses a lot of formulas — that's no lie. You'll be scuppered if you start trying to answer a question without the proper formula to start you off. Thankfully, CGP is here to explain all things formula-related.

You're Given these Formulas

Fortunately, those lovely examiners give you some of the formulas you need to use.

> **For a sphere with radius *r*, or a cone with base radius *r*, slant height *l* and vertical height *h*:**
>
> **Volume of sphere** $= \frac{4}{3}\pi r^3$ **Volume of cone** $= \frac{1}{3}\pi r^2 h$
>
> **Surface area of sphere** $= 4\pi r^2$ **Curved surface area of cone** $= \pi r l$

And, actually, that's your lot I'm afraid. As for the rest...

Learn All The Other Formulas

Sadly, there are a load of formulas which you're expected to be able to remember straight out of your head. Basically, any formulas in this book that aren't in the box above, you need to learn. There isn't space to write them all out below, but here are the highlights:

Area of parallelogram = base × vertical height

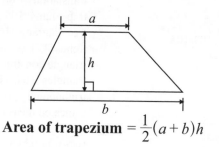

Area of trapezium $= \frac{1}{2}(a+b)h$

Exterior angle of regular polygon $= \frac{360°}{n}$

Sum of interior angles of any polygon $= (n-2) \times 180°$

where *n* is the number of sides

For a circle with radius *r*:
Circumference $= 2\pi r$
Area $= \pi r^2$

Volume of prism = cross-sectional area × length

Volume of cylinder = area of circle × height $= \pi r^2 h$

Surface area of cylinder $= 2\pi r h + 2\pi r^2$

Compound Growth and Decay:
$N = N_0(\text{multiplier})^n$

Area of sector $= \frac{x°}{360°} \times$ Area of full circle

Length of Arc $= \frac{x°}{360°} \times$ Circumference of full circle

For a right-angled triangle:
Pythagoras' theorem: $a^2 + b^2 = c^2$
Trigonometry ratios:
$\sin x = \frac{O}{H}$, $\cos x = \frac{A}{H}$, $\tan x = \frac{O}{A}$

Compound Measures:
$\text{Speed} = \frac{\text{Distance}}{\text{Time}}$ $\text{Density} = \frac{\text{Mass}}{\text{Volume}}$ $\text{Pressure} = \frac{\text{Force}}{\text{Area}}$